This is a continuation in the series of publications produced by the Center for Advanced Concepts and Technology (ACT), which was created as a "skunk works" with funding provided by the CCRP under the auspices of the Assistant Secretary of Defense (NII). This program has demonstrated the importance of having a research program focused on the national security implications of the Information Age. It develops the theoretical foundations to provide DoD with information superiority and highlights the importance of active outreach and dissemination initiatives designed to acquaint senior military personnel and civilians with these emerging issues. The CCRP Publication Series is a key element of this effort.

Check our Web site for the latest CCRP activities and publications.

www.dodccrp.org

DoD Command and Control Research Program

Assistant Secretary of Defense (NII)
&
Chief Information Officer
John G. Grimes

Principal Deputy Assistant Secretary of Defense (NII)
Dr. Linton Wells, II

Special Assistant to the ASD(NII)
&
Director of Research
Dr. David S. Alberts

Library of Congress Cataloging-in-Publication Data

Alberts, David S. (David Stephen), 1942-
Understanding command and control / David S. Alberts, Richard E. Hayes.
p. cm. -- (Future of command and control)
Includes bibliographical references and index.
ISBN 1-893723-17-8
1. Command and control systems. I. Hayes, Richard E., 1942- II. Title. III. Series.
Cover and illustrations by Joseph Lewis
UB212.A435 2006
355.3'3041--dc22

2006000771

2006

THE FUTURE OF COMMAND AND CONTROL

UNDERSTANDING COMMAND AND CONTROL

David S. Alberts

Richard E. Hayes

Table of Contents

List of Figures

ACKNOWLEDGMENTS

We would like to thank the many individuals and organizations who supported and assisted in the writing of this book. Their valuable ideas, comments, and contributions enriched our own thoughts and arguments.

Many of the ideas presented here were the subject of an intense 3-year international research collaboration conducted under a charter from NATO's Research and Technology Organisation (Studies, Analysis, and Simulation Panel Working Group SAS-050),[1] which produced a comprehensive C2 Conceptual Reference Model. We benefited greatly from participating in that activity.

[1] The members of SAS-050 included: Dr. David Alberts (US), Mr. Graham Cookman (UK), Mr. Natalino Dazzi (IT), Dr. Lorraine Dodd (UK), Ms. Petra Eggenhofer (GE), Mr. Geir Enemo (NO), Mr. Fernando Freire (PO), Dr. Anne-Marie Grisogono (Australia), Dr. Richard Hayes (US), Dr. Gary Horne (US), Dr. Reiner Huber (GE), Mr. Reinhard Hutter (GE), Mr. Gert Jensen (DK), Ms. Sarah Johnson (US), Mr. Nickolas Lambert (NL), Mr. Viggo Lemche (DK), Ms. Danielle Martin (US), Mr. Graham Mathieson (UK), Dr. Daniel Maxwell (US), Dr. James Moffat (UK), Mr. Allen Murashige (US), Mr. Klaus Niemeyer (GE), Mr. Arne Norlander (SE), Maj. Paulo Nunes (PO), Dr. Paul Phister (US), Mr. Valdur Pille (CA), Mr. Dieter Rathmann (GE), Mr. Xander Roels (NL), CPT Jens Roemer (GE), Mr. Gunther Schwarz (GE), Mr. Mark Sinclair (US), M.Sc. Mink Spaans (NL), Ms. Kristi Sugarman (US), LTC (Ret) Klaus Titze (GE), Mr. Rick van der Kleij (NL)

We were also fortunate to have been able to draw upon the insights and experiences of many colleagues. We would like to thank five individuals in particular for reviewing the manuscript for this book and providing us with thoughtful and detailed peer reviews. These individuals are Graham Mathieson, Professor James Moffat, Dr. David Noble, Dr. David Signori, and Dr. Ed Smith. The willingness of these senior professionals to take the time necessary to offer constructive criticisms is greatly appreciated.

In addition, the members of the Information Age Metrics Working Group (IAMWG), senior personnel who assemble monthly to look at important issues, gave us rich feedback on the early draft material and participated in collegial discussions of key topics covered in the book. Regular members of the group include Dr. Ed Smith, John Poirier, Dennis Popiela, Dr. Mike Bell, Mark Sinclair, Dr. Mark Mandeles, Julia Loughran, Kirsch Jones, Eugene Visco, Dr. Larry Wiener, Manual Miranda, Pat Curry, Donald Owen, Mitzi Wertheim, RADM Evelyn Fields (ret.), and Dr. Paul Hiniker.

We were also very ably supported by Joseph Lewis, who provided the technical edit, created the graphics and cover artwork, and did the key layout work. In those processes, he made valuable suggestions that helped us make the text clearer and better organized. Margita Rushing managed the publication process with her usual dedication and efficiency.

Preface

Thomas Kuhn observed that progress in science is not linear but that it exhibits periods "of peaceful interludes punctuated by intellectually violent revolutions."[2] These revolutions are what he called *paradigm shifts.* The world of Command and Control is in the midst of a paradigm shift, a change in the way we think about the subject. After years of trying in vain to make what historically has become known as Command and Control work in an era of complex coalition civil-military operations, there is an increasing willingness to rethink the subject. At the same time, Information Age concepts and technologies offer opportunities to do things we could never do before. The "stars are aligning," matching our *need* to change with the *means* to change. Therefore, it is time to move on. It is time to recognize that, if we are to be successful in meeting the 21st century challenges that we face, there will be major discontinuities between the Command and Control concepts and practices being taught and practiced today and those of tomorrow.

Understanding Command and Control is the first in a new series of CCRP Publications that will explore the future of Command and Control. A major discontinuity that will need to be

---

[2] Kuhn, Thomas. *The Structure of Scientific Revolutions.* Chicago, IL: University of Chicago Press. 1996. p. 10.

addressed will be the definition of the words themselves. This is because the way that these words have been defined drastically limits the available solution space and points us in the wrong direction. This creates major problems for both authors and readers. It makes it very difficult to communicate effectively in a medium that is half duplex, where there is no ability for the authors and the readers to interact in real time; for readers to express their questions and concerns and for the authors to clarify and explain. Recognizing this, there was and continues to be a great deal of discussion about what to call this first book in the series and the functions it discusses. We chose to continue to use the term *Command and Control* despite its obvious problems because we wanted to find the appropriate audience, those who are interested in Command and Control, even if what they mean by these terms is very different from how we believe we should be thinking about the subject.

This book begins at the beginning: focusing on the problem(s) Command and Control was designed (and has evolved) to solve. It is only by changing the focus from *what* Command and Control is to *why* Command and Control is that we will place ourselves in a position to move on.

Various CCRP Publications have foreshadowed this need to break with tradition. *Coalition Command and Control* (Mauer, 1994) raised fundamental questions about how to re-interpret Command and Control in the context of a coalition. *Command Arrangements for Peace Operations* (Alberts and Hayes, 1995) suggested some answers and raised some additional questions. *Coping with the Bounds* (Czerwinski, 1998) addressed the challenges associated with complexity. *Network Centric Warfare* (Alberts, Garstka, and Stein, 1999), by focusing on shared awareness and self-synchronization, set the stage for *Power to the Edge: Command and Control in the Information Age* (Alberts and Hayes, 2003). Each of these publications has, in its own way, contributed to the ongoing exploration of ways to improve Command and Control. However, virtually all of this exploration has occurred in close proximity to the status quo.

Although transformation, which is inarguably about disruptive innovation, is a major policy objective of not only the DoD, but militaries throughout the world, these commitments to transformation have yet to shift the focus of Command and Control analysis and experimentation to the other side of the discontinuity. This series is meant to stimulate and contribute to the exploration of the other side.

David S. Alberts

Washington, DC

January 2006

CHAPTER 1

INTRODUCTION

Understanding Command and Control (C2) is no longer an option; it is a requirement. This introductory chapter will explain why, if we want to make significant progress on Defense transformation or succeed in 21st century operations, we need to understand Command and Control thoroughly. This book is intended to provide a sound foundation for efforts to better understand Command and Control.

IMPORTANCE OF UNDERSTANDING C2

The mission challenges of the 21st century have increased significantly. Fortunately, new concepts of operations and approaches to Command and Control are able to provide significantly increased capabilities to deal with these challenges.

Today's missions differ from traditional military missions, not just at the margins, but qualitatively. Today's missions are simultaneously more complex and more dynamic, requiring the collective capabilities and efforts of many organizations in order to succeed. This requirement for assembling a diverse set of capabilities and organizations into an effective coalition is accompanied by shrinking windows of response opportunity. Traditional approaches to Command and Control are not up to the challenge. Simply stated, they lack the agility required in the 21st century.[3]

Fortunately, advances in information technologies have created a new space within which individuals and organizations can operate. Those individuals and organizations that have learned to take advantage of the opportunities afforded by operating in this new space have realized a significant competitive advantage over those that have ignored these opportunities. The Department of Defense (DoD) has recognized that these opportunities exist and is committed to an Information Age transformation. This transformation has two major axes: one focused on understanding 21st century mission challenges and one focused on Network Centric Operations (NCO) (and DoD business processes).

The network-centric axis of transformation is anchored by the tenets of Network Centric Warfare (NCW) and Power to the Edge principles. At the risk of oversimplification, NCW is a two-step process: first, achieving shared awareness, and second, leveraging shared awareness to achieve a greater degree of self-synchronization, leading to dramatic increases in both agility and effectiveness. The magic of NCW is the emergence of self-synchronizing behavior.[4] Ultimately, the most important contribution that network-centric approaches to C2 will make is increasing force or enterprise agility. This is because

---

[3] This is discussed in some detail in: Alberts and Hayes, *Power to the Edge*. Washington, DC: CCRP Publication Series. 2003.

[4] See: Alberts et al., *Network Centric Warfare*. Washington, DC: CCRP Publication Series. 1999. p. 175.

the mission challenges of the 21st century place a premium on being agile.

Like many of our coalition partners, DoD has invested in building a robust, secure, ubiquitous infostructure and, as a result, the coming years will see greatly increased connectivity, quality of service, and interoperability. DoD has adopted Power to the Edge principles. The early manifestation of these has been in DoD's Data Strategy to facilitate and encourage widespread information sharing and collaboration. These steps will move us toward shared awareness, but they are not sufficient to help us leverage shared awareness. To take this second step, we need to move from a networked infostructure to create a networked or Edge organization. To accomplish this, we need to develop new approaches to Command and Control. These include the creation of robust socio-technical networks that rely upon human behaviors that are facilitated and supported by technical means.

Therefore, new C2 Approaches are the fulcrum of an Information Age transformation of the DoD and understanding Command and Control is among the most important and urgent tasks we have on the critical path to transformation and the ability to meet 21st century mission challenges.

Purpose

The purpose of this book is to provide the conceptual foundation for the C2 research and experimentation necessary to develop and explore the new C2 Approaches needed for this transformation. In developing and presenting this foundation, we are unwaveringly focused on the future, not the past. While the foundation presented can be used to understand traditional approaches to C2, its value lies in its ability to help us understand new network-centric approaches. Our intended audience is very broad because, without a broad-based understanding of C2, progress is problematical. At the same time, we are also addressing issues crucial to the C2 community, from practitioners to theoreticians.

Organization of the Book

This book begins with a short "Reader Orientation" intended to stress key issues that differentiate this work from previous thinking on the topic. We then turn to exploring what it means to "understand" something, varying degrees of understanding, and the implications of understanding to different degrees. This also includes a discussion of models, with a focus on what a conceptual reference model—the instantiation of a model—is and the differences between a value view and a process view. This is followed by an introduction to the concepts of Command and Control, starting with why Command and Control is needed and the functions that need to be accomplished to achieve its purposes.

The discussion of C2 then moves to the nature of the C2 Approach space, a space that contains the full range of options available to us for accomplishing the functions of command and the functions of control. There is a set of functions, like inspiration, that is often associated with Command and Control because it is a property of commanders, not a property of a C2 Approach. Furthermore, many if not all of the functions that we associate with Command and Control need to be performed by an individual or group—they instead may be emergent properties that arise within an organization.

At this point, we turn our attention to presenting a C2 Conceptual Model. An overview of this model is followed by in-depth treatments of the C2 Approach, the C2 value chain, C2 process views with examples, and influences that affect the values of key C2-related variables and the relationships among them. The concluding discussion identifies the critical path to developing a better understanding of Command and Control.

CHAPTER 2

READER ORIENTATION

Many readers will find this book challenging. As a subject, Command and Control has a reputation for being arcane, even among individuals who arguably are or have been practitioners. The words Command and Control individually and collectively mean different things to different communities. As explained in the Preface to this book, we chose, for the moment, to stay with these words even though we believe that the way they have been defined and understood limits our ability to accomplish the functions that Command and Control seeks to accomplish.

To expect that anyone will come to this book without some preconceived notion of what the terms mean is unrealistic. But we want readers who are interested in Command and Control to think about what we have to say. Our hope is that readers will be able to, at least for a while, put aside what they "know" about Command and Control and approach our treatment of the subject with an open mind. Having understood what we are proposing, readers are of course free to accept, argue about, or help us to improve these concepts.

We offer the following conceptual trail markers that we hope will assist readers in orienting themselves for this book's journey. We will employ a special font to remind readers that we are talking about our concept of **Command and Control** rather than traditional definitions or uses.

- **Command** and **Control** are separate but interrelated functions.
- **Command and Control** involves only the specific functions we explicitly associate with these terms. Thus, **C2** is not about "who"; it is about "what."
- **Command and Control** does not encompass all of the decisions made by individuals or organizations nor all of the decisions that emerge from collective behavior; only the ones directly associated with the functions of **C2**.
- **Command and Control** applies to endeavors undertaken by collections of individuals and organizations of vastly different characteristics and sizes for many different purposes.
- The most interesting and challenging endeavors are those that involve a collection of military and civilian sovereign entities with overlapping interests that can best be met by sharing information and collaboration that cuts across the boundaries of the individual entities.
- **Command and Control** determines the bounds within which behavior(s) are to take place, not the specific behaviors themselves. The degrees of freedom associated with these bounds can vary greatly.
- Thus, **C2** establishes the conditions under which sensemaking and execution take place. **C2** is separate from sensemaking and its operational implementations.
- It is important to always keep in mind that there are many different approaches to accomplishing these func-

tions. No specific approach or set of approaches defines what **Command and Control** means.

- **Command** and **Control** are fractal concepts. They can be applied to all subsets of an enterprise; to the functions performed; to the levels of the organizations; to the focus of the activity, whether strategic or tactical. Membership in these fractals may overlap with individual entities and groups belonging to multiple fractals dynamically.
- Different **Command and Control** Approaches will be appropriate for different sets of purposes or circumstances.
- Different **Command and Control** Approaches may be taken by different sets of entities in an enterprise, and may change over time.
- Successfully accomplishing the functions of **Command and Control** *does not* necessarily require:
  - Unity of command (an individual in charge)
  - Unity of intent (an intersection of goals)
  - Hierarchical organizations
  - Explicit control

The effect of these conceptual trail markers is to take those who choose to explore **Command and Control** on a journey of discovery, unconstrained by existing notions and practices. This is meant to ensure that those who sign up for this journey will be operating outside of their comfort zones in the hope that the full range of possibilities will be explored.

Chapter 3

Understanding

Nature of understanding

There are many different ways to explain the concept of *understanding*, each with its own nuances. To first order, *to understand* something is to be able to grasp its nature or significance; *to understand* is to comprehend (an idea or a situation); *to understand* is the ability to offer an explanation of the causes of an observable state or behavior. In our past work, we have stressed that "understanding" goes beyond knowing what exists and what is happening to include perceptions of cause and effect, as well as temporal dynamics.

Since the dawn of empiricism,[5] understanding has been associated with systematic observation, experience, and expertise rather than revelation. We say that we understand something when the result seems reasonable to us and we say that we do

[5] The origins of empiricism are usually traced back to the 17th century and Galileo, relatively recent in terms of civilization.

not understand it when the result is unexpected or (at least to us) without a logical explanation.

Understanding resides in the cognitive domain[6] and, like everything in the minds of humans, is subjective, influenced by perceptual filters and biases. However, one's understanding may not be "correct," that is, it may not conform to objective reality.[7] Thus, one can apply attributes to understanding that correspond to the attributes we associate with information, including correctness and completeness.[8]

To understand something does not mean that one can predict a behavior or an event. Prediction requires more than understanding, thus even if one understands a phenomenon, one may not be able to predict, with anything that approaches a level of usefulness, the effect(s) of that phenomenon. Prediction requires actionable knowledge, specifically the values of the variables that determine (or influence) the outcome in question.[9] Operationally, the most that can be expected is to identify meaningfully different alternative futures and indicators that those alternatives are becoming more or less likely over time.

---

[6] The other domains in the models discussed in CCRP Publications include the physical domain, the information domain, and the social domain. Some areas of study break these domains into further subdivisions.

[7] For the purposes of these discussions, we assume that there is in fact a reality that exists outside of human minds, a reality that can be observed and characterized. Included in this reality are the perceptions and understandings of other entities.

[8] The attributes that we associate with the quality of information and understanding are discussed in our treatment of the conceptual model.

[9] The nature of understanding encompasses the knowledge about the relationships among the variables in question, thus all we need in order to determine the result are their values.

Understanding is also insufficient to improve a situation. Improvement that is deliberate and not the result of trial and error requires both the ability to predict and the ability to control the values of some or all of the variables that affect the outcome. Thus, the value or utility of understanding in order to improve a situation depends upon specific knowledge and the degree to which one can control or influence key variables.

Degrees of understanding

There are degrees of understanding that correspond to a scale that runs from a cursory understanding to a complete understanding. In terms of understanding **Command and Control**, a cursory understanding of **C2** would involve understanding only what **C2** is trying to accomplish, that is, the result that **C2** is designed to achieve. A greater degree of understanding requires recognition of the different **C2** Approaches and their applicability. The degree to which one can answer the following questions about **C2** corresponds to the degree to which one understands its nature and its application to selected situations.

- What are the possible **Command and Control** Approaches? (how desired results could be accomplished)
- What are the key differences among **Command and Control** Approaches? (the dimensionality of the **C2** space)
- What influences the ability of a **C2** Approach to realize its objectives?
- Which approaches are appropriate for a given set of circumstances?
- What can be expected if a particular approach is adopted and a specific set of circumstances is obtained?

Despite the fact that military organizations have practiced **Command and Control** for millennia, the answer to even the first of these (i.e., possible approaches) is not definitively known because military organizations have, until very recently, only explored a small subset of the approaches[10] that appear to have potential.

This book seeks to provide a conceptual foundation that can be used to develop a better understanding of **Command and Control** so that answers to these questions can be found. One of the biggest problems is that there has been relatively little effort expended on finding answers to some of these questions because of a prevailing view that we have a **C2** Approach that works well (or that it is thought to have served us well so far). In fact, the view that traditional **Command and Control** Approaches have worked well is debatable and the view that traditional approaches will continue to serve us well is not supported by current events and operations. The relevant threats, operating environments, technologies available, and our understanding of human enterprises are all changing.

FACTS, THEORIES, AND MODELS

If we are to improve our understanding of **Command and Control**, then we will need to establish facts, develop testable theories, and instantiate these theories in models. In short, we must build a body of knowledge, gain experience, and develop expertise. To accomplish this, we need to observe reality, intellectually develop conceptual models, and design and conduct

---

[10] Traditional views on **C2** are very strongly held. Thus, some of the **C2** Approaches that we will consider may not be considered "**C2**" by some. Therefore, we have chosen to use "approaches to accomplishing the functions associated with **C2**" to make it clear that we are looking beyond what has been done to date.

Facts, theories, and models

experiments to calibrate and validate these models. This entails the collection of empirical evidence, the conduct of analyses, the publication of results, and the archiving of data. These tasks are iterative.

A complete set of facts is not necessary to formulate a theory or construct a model. Theories and models are most often a mix of what we know (or think we know) and what we think (conjecture or hypothesize). Theories are almost always conceived from a limited understanding (having only a fraction of the necessary facts and the relationships among them) and serve to focus our efforts to identify additional relevant variables and to discover relationships. A fact is a piece of information having objective reality, and facts reside in the information domain, but how individuals and groups interpret facts is another thing. These interpretations or perceptions occur and reside in the cognitive domain. Therefore, theories that address human behavior must deal with both facts and the ways in which they are perceived.

A theory[11] consists of the abstract principles of a body of facts. The dictionary notes that the term *theory* can be applied to both a science and an art (as in music theory). Given that many think of **Command and Control** as an art and a science,[12] the notion of a "theory of **C2**" would be appropriate in either case. A theory and a model that instantiates a theory consist of a set of facts (or assumptions) and the relationships among them. A theory or model can be as simple as the economic price theory we learned in Economics 101: [P = *f* (S,

---

[11] Merriam-Webster Online Dictionary, "theory." <http://www.m-w.com/> March 2006.

[12] With command being an art and control a science. We think this formulation is too simplistic and hence misleading.

D)], where P = price, S = supply, and D = demand. However, operationalizing this theory is far from simple and has occupied many economists for a long time.

Theories and the models that instantiate them are representations for a purpose. We were first introduced to models as children. Dolls, toy cars, guns, and swords are iconic models (physical representations of the real thing). Iconic models are also used extensively to test designs for ships and airplanes in tanks and wind tunnels. This allows us, at relatively low cost, to subject these designs to various conditions and observe their behaviors. It allows us to go beyond what we normally experience in the real world and test these designs under extreme conditions. Additionally, iconic models need not be complete representations of reality; they only need to provide an adequate representation of the characteristics that we are interested in for the purpose of the experiment. In the tank or wind tunnel, this may be only the shape of a hull or fuselage.

Iconic models are relatively easy to build and they are easy to relate to the theory or object they are designed to represent, but they are not easy to change. If one wants to explore a series of hull shapes that represent changes to a particular parameter, then many models need to be built, each one representing a different value of the parameter in question. Testing the effect of the value of the parameter in question thus involves running a series of tests on each model. One can see how working only with physical models might be a very time consuming process and hence limit exploration of a parameter. Exploring multiple interrelated parameters would be even more cumbersome.

Facts, theories, and models

Different kinds of models are better suited for simultaneous exploration of the effects of a number of parameters. These more agile[13] models are mathematical models and simulation models, both of which are instantiations of a conceptual model. These models are designed to allow for the changing of the values of a parameter or a set of parameters and then determining the effect that this has on the variables of interest.

Conceptual models are representations of how we think (conceive) about something, in this case **Command and Control**. The building blocks of these models are concepts, which translate into one or more variables and the relationships among them. The degree of specificity with which these relationships are expressed in conceptual models varies from the existence of a relationship or influence to a more definitive expression of the nature of the relationship. Conceptual models are often depicted graphically with the concepts expressed as boxes or other shapes and the relationships between and among the concepts as lines or directional arrows. Mathematical models consist of sets of related equations.

Conceptual models, mathematical models, and simulation models all have the same basic building blocks: variables and the relationships among them. Conceptual models and mathematical models are not working expressions, while simulation models are, in reality, tools that bring conceptual models or mathematical models to life, producing outputs from a set of inputs. Other tools serve this purpose as well, including dynamic and linear programming, expert systems, and the

---

[13] The use of the word *agile* is appropriate because these models' characteristics correspond to many if not all of the dimensions of agility that are defined in *Power to the Edge* and are included in the NCO Conceptual Framework and the NATO **C2** Conceptual Model. Alberts and Hayes, *Power to the Edge.* pp. 123-159.

familiar but versatile spreadsheet. A major difference among these types of tools is whether they are event-oriented, rule-oriented, or instantiating formulas. Simulation models sometimes do all three, generating events on a predetermined or stochastic basis, having agents that employ rules that govern decisions, and calculating the values of parameters using static or dynamic formulas.

Building a conceptual model

Despite the ease of constructing a conceptual model, that is, going to a whiteboard and drawing a bunch of shapes and lines, building a *meaningful* conceptual model is quite difficult. The most important decisions involve what to include and what not to include. When a piece of Mozart's was criticized for having "too many notes," the composer replied that the piece did not have too many or too few notes but exactly the right number of notes. So too does a model that is "fit for use." The important thing to consider is whether or not the model serves its intended purpose.

Well-conceived and constructed models do not have too many or too few variables, but just the right ones. The number of variables must be kept to the bare minimum needed in order to enable the model to communicate its concepts to others. For this reason, less is more. Keeping the number of variables and relationships under control makes the model as simple as possible and thus as easy to understand as possible. This requires the model to extract the essence of reality, and only the essence. The way in which designers of a conceptual model balance the need for simplicity with the need for fidelity often determines success. One way of dealing with these conflicting objectives is to have a number of depictions or views of the conceptual model, each of which serves a specific purpose and a way of organizing detail.

Identifying the minimum essential concepts

To illustrate this point and the nature of a conceptual model, we have built a relatively simple model designed to explore the control[14] of a room's temperature. Figure 1 depicts a set of the minimum essential concepts (one or more variables that represent the necessary elements of the model) needed to explore the approach taken in an attempt to keep the temperature of a room within desired bounds.

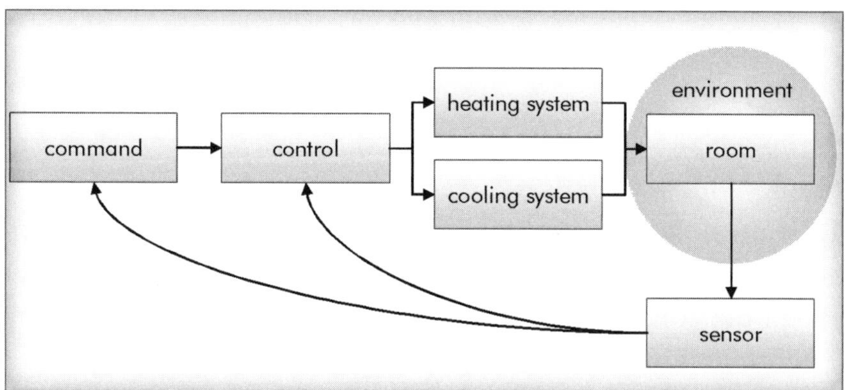

FIGURE 1. MINIMUM ESSENTIAL CONCEPTUAL ELEMENTS FOR ROOM TEMPERATURE CONTROL MODEL

The minimum essential elements include: command, control, two sets of capabilities (heating and cooling), the target (room), its environment, and a sensor (thermometer). In its most common instantiations, command involves the determination of

[14] The word *control* is used here in its traditional sense: a set of actions taken to create an effect or achieve a desired outcome. Later in this book, we will offer a more restrictive definition of control in the context of **Command and Control**.

the desired temperature and the function of control is actually built into a rather simple thermostat. The function of control translates the desired temperature into a set of rules that govern what actions are taken. As we will later define these terms, the thermostat embodies elements of both control and sensemaking. If the temperature behavior in the room does not meet expectations, a number of actions can be taken. Command may decide to reset the desired temperature, buy a new or different thermostat (hence changing the nature of control), modify, repair, or replace one or both of the systems (hence, for example, reallocating resources), just to name a few of the possibilities that create a different set of conditions. This simple model can represent a wide variety of **Command and Control** Approaches, help us to understand the nature of the task involved, and inform a wide range of decisions.

Instantiation of concepts in a mathematical model

How much we need to know about any one of the model's conceptual elements depends on (1) the nature of the purpose or use of the conceptual model and (2) reality. For example, if the environment were invariant, then the effect of the environment on the room temperature could be accounted for as part of the characteristics of the room and for all intents and purposes would not need to be depicted separately. If the characteristics of the room were also invariant over time, then all we would need to represent the concept of the room would be two functions: one for heat loss over time, the other for heat gain over time, both of which would be conditional on the current room temperature. If the sensor reported room temperature accurately, instantaneously, and with a precision that was appropriate, then we would not need to represent the sensor.

In a similar fashion, the heating and cooling systems can be represented by mathematical expressions that are functions of current temperature and time since the heater/air conditioner was turned on. Combining these mathematical expressions for the systems and the room characteristics into a single expression is rather straightforward. Thus, room temperature at time t may be expressed as $T(t) = T(t-\Delta t) + \partial(t-\Delta t)$, where ∂, the gain/loss of room temperature between $t-\Delta t$ and t, is a function of the ability of the heater/air conditioner to warm/cool the room at time $t-\Delta t$, the room temperature at time $t-\Delta t$, and both room and environmental conditions.

We now turn our attention to the nature of the **Command and Control** Approach that is being considered. First, as we earlier assumed, the function of command consists of picking a desired temperature for the room and the approach to control consists of a translation of this intent into a simple decision, namely to turn the heater on when the room temperature falls to a predetermined temperature, turn the heater off when the temperature is at or above a predetermined temperature, turn the air conditioner on when the room temperature rises to a predetermined temperature, and turn the air conditioner off when the room temperature is at or below a predetermined temperature. The function of control is to select these predetermined temperatures.

Figure 2 instantiates these assumptions in the form of a mathematical model. This mathematical model is deterministic, that is, the behavior of room temperature is totally determined by the values of the parameters embedded in the temperature gain/loss function.

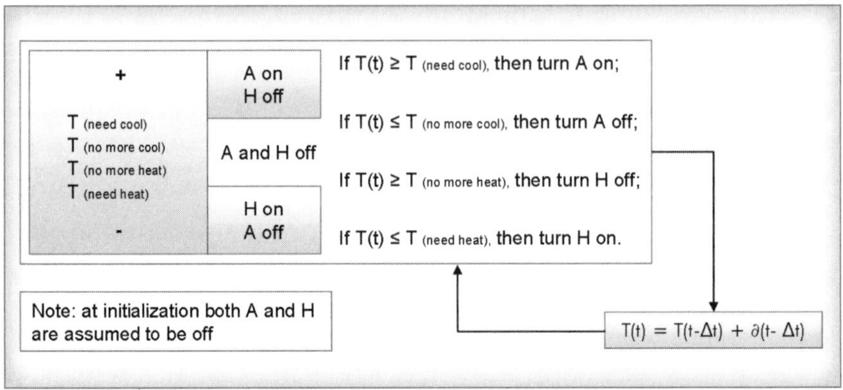

FIGURE 2. MATHEMATICAL MODEL: TEMPERATURE CONTROL[15]

This very simple model can be used to investigate a number of important questions. Given the assumptions: the room and environmental characteristics, the performance characteristics of the heating/cooling systems, and the **Command and Control** Approach,

- How much of the time can the room temperature be maintained at x degrees plus or minus y degrees?
- If the characteristics of the environment were to change (specify the change), what effect would it have on the ability to maintain a given room temperature?

The first of these questions involves merely plugging in the value of the target temperature and calculating room temperature as a function of time. The second question involves a revision to the formula, $T(t) = T(t-\Delta t) + \partial(t-\Delta t)$, in effect, a modification to the mathematical model. Note that no revisions to the conceptual model are required.

---

[15] We would like to thank Dr. Richard Daehler-Wilking of SPAWAR for assisting our formulation of this diagram.

The way in which command has been represented in this example equates to *command by intent*, a statement of the desired outcome. The approach to control is interventionist,[16] that is, specific orders are given to the "forces" (the heater and the air conditioner) at irregular intervals when the temperature reaches x (a specified event or condition). Thus, control here consists of a simple decision linked to scripted behavior. These types of decisions can be easily automated (with a thermostat, they are built in as hardware, or with a more sophisticated thermostat, a combination of hardware and software).

One conceptual model, many instantiations

While the above instantiation of the conceptual model is suitable for the purposes defined above (providing answers to the questions), it may not be suitable for addressing a different set of questions. However, as long as the conceptual model is suitable, the mathematical model may be altered to reflect changes in assumptions. For example, the nature of command intent could be altered from a simple target temperature to a target temperature that changes with time or circumstances. For example, the room could be maintained at a specific temperature during working hours and a different temperature the rest of the time. A more sophisticated behavior may be desired, for example, a temperature target for the room when occupied and a different target for the room when empty. This would require modeling a sensor that was capable of detecting whether the room was occupied or not, but the basic structure of the original conceptual model would be the same.

---

[16] For a discussion of the interventionist approach and others, see: Alberts and Hayes, *Power to the Edge.* pp. 18-27.

There are, of course, ways to alter the temperature of a room other than simply turning a heater or air conditioner on or off. For example, windows and doors could be opened and closed, shades raised or lowered, and lights turned on or off. In addition, the number of people occupying the room could alter its temperature.

To this point, the model has focused only on the variables found in the physical and information domains. The physical domain is the source of the equations regarding temperature changes; the information domain consists of reports of room temperature.

Now let us consider an alternative way of defining the room conditions that we seek (desired values). Instead of using a measure that can be determined by physical measurement, like room temperature, one could use the comfort level of the room's occupants. While there is a relationship between room temperature and comfort level, comfort level will vary from person to person and with other factors including humidity, air movement, and light levels. They will also be affected by apparel choices.

> Will the need to consider means other than heating and cooling of the room's air (e.g., to affect humidity or light levels) affect the way we need to formulate our conceptual model?

If we look at the conceptual model, we see that the *means* identified to change room temperature includes a heating system and a cooling system. Generally, we would interpret a heating system to consist of an oil, gas, or electric furnace with air ducts, or perhaps a radiant system with steam or hot water or a heat pump. There are of course other possibilities that might leap to mind such as an active or passive solar heating system. If we liberally interpret the terms, a heating and cooling system may be powered in any of a variety of ways and also include the ability to change airflow, light conditions, humidity, and the degree to which the room "membrane" isolates it from the environment. If we broadly interpret these terms, the conceptual model, as we have formulated it, remains appropriate for our purposes as far as the way we can alter the physical characteristics of the inside of the room.

> Will the adoption of "comfort level" as the measure of desired room conditions require us to change the conceptual model?

The conceptual model includes the existence of a sensor that reports on the condition of the room. In our initial discussion of the problem, we interpreted this to mean a thermometer. If there was such a thing as a "comfort sensor," we could simply replace the temperature sensor with a comfort sensor, but of course there is no such generally accepted device. At this point, we have three options. The first is to include a set of sensors that measure a variety of physical factors that are known to affect an individual's comfort level to obtain an indicant of comfort. The second is to indirectly measure the comfort level of individuals occupying the room by their behavior (e.g., sweating or teeth chattering). The third is to have the occupants directly report their level of comfort. In each of these cases, all we have done is to specify what we mean by "sensor" and thus the conceptual model, as formulated, remains appropriate.

> How will the adoption of "comfort level" change the way we look at **Command and Control**?

If we use a set of measurable physical characteristics as an indicator of comfort, we need only (1) change the way command intent is expressed (i.e., target values for a defined function of the set of variables), (2) adjust the simple decision accordingly, and (3) map it to a new set of actions related to the means of affecting the characteristics of the inside of the room (e.g., lowering a shade). Again, this does not require a change in the way we have formulated the conceptual model.

Dealing with each of these changes to the way that we think about how we affect the conditions inside the room and how we value the outcome we have achieved does not require a change to our conceptual model. However, they do require us to identify new variables and relationships and the introduction of the value metric of "comfort" requires considerations that involve the cognitive domain.

Let us now consider a more radical way of controlling room comfort. Suppose instead of trying to sense the comfort level of individuals occupying the room and taking appropriate action, we provide them with the means to do it themselves. With access to the appropriate means, occupants can adjust the heating and cooling systems, adjust air movement using fans or room openings, change humidity levels, and adjust lighting conditions. Adopting this approach clearly decentralizes control. The decentralization of control can lead to actions being taken that may conflict with one another. Therefore, we need to deal with the interesting question of how two or more occupants could accomplish the functions associated with command. However they decide to do it (delegate, negotiate, collaborate), the interactions between and among the occupants would be in the social domain.

One conceptual model, many instantiations

Once again there is no need to revise the conceptual model, but once again new variables and relationships need to be introduced. We have come a considerable way from our original notion of a room's temperature being controlled by what amounts to a thermostat set to a given level. Some of the complexity that we added (new variables and relationships) was a result of a change in the intended use of the model, some by a recognition that the assumptions were not appropriate, some by changing the way we define the room conditions we seek, and some by altering our **Command and Control** Approach.

Process versus value views

At this point, the model formulation addresses all four domains: physical, information, cognitive, and social. It can be used to describe the conditions in the room over time and compare these room conditions to a specific target condition or set of conditions. As formulated, this model has only one embedded measure that we can use to characterize success: the nature of the difference between a desired outcome and the actual outcome. The reason that the model contains this measure is related to process rather than value. This comparison is required for the processes of **Command and Control**. The target conditions are an expression of the output of the process we call *command* and an input to the process we call *control*. The other input to control in the model is a reading of room conditions.

This model, as formulated, can be used to explore different heating and cooling methods and to see what their effects are on the ability to keep room conditions in sync with a dynamic target. This model can also be used to determine what range of environmental conditions can be tolerated.

Suppose that we wanted to know:

- How much energy does it take to keep the temperature within specified bounds? or
- Does it make more sense to better insulate the room or increase the capacity (effectiveness) of the heating/cooling system?

The first question requires some additional information about how much of the time the heater is on and how much of the time the air conditioner is on, and how these relate to a measure of energy consumption. Energy consumption can, in turn, be related to cost. The second question requires an understanding of the costs and system performance improvements associated with each of the two ways of improving our ability to maintain room conditions.

In general, these issues assume that there are things we can control and that our choices make a difference in our ability to create the effects or achieve the goals specified by command intent. Beyond our ability simply to realize intent, there are issues of efficiency and/or cost associated with our choices. There is an implied value chain that links the quality of command intent and its expression to system performance and on to measures of value associated with outcomes. Information quality clearly affects each of these. Making the value chain explicit helps us to focus on what really matters rather than on describing behaviors.

Figure 3 identifies the value metrics associated with each of the concepts in the conceptual model.

One conceptual model, many instantiations

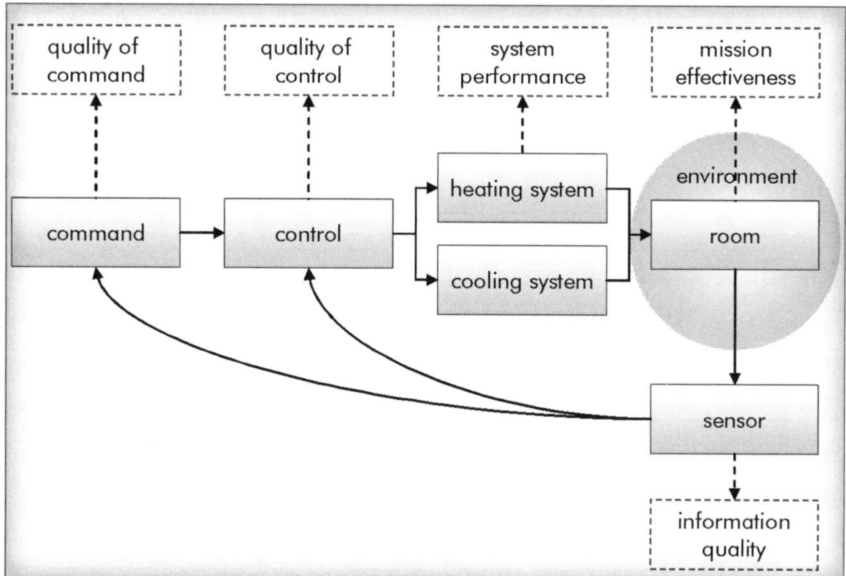

FIGURE 3. CONCEPTUAL MODEL WITH VALUE METRICS

Value metrics, like concepts, can consist of one or more variables. For example, the quality of the information provided by a temperature sensor is determined by, at a minimum, its accuracy, currency, and precision. A meat thermometer that provides you with an instantaneous readout that correctly tells you that the meat is well-done, medium, or rare may or may not be as useful as one that tells you, within one degree of accuracy, that the meat was 140 degrees 30 seconds ago. The issue at hand is "fitness for use" and that in turn may depend on the nature of the situation (e.g., rate of temperature change, or the degree of education and experience of the cook).

The inclusion of value concepts, in addition to process concepts, opens the door to a richer set of uses for the model. As the above example shows, if we get the basic concepts right, we will be able to develop a series of instantiations that helps us to deal with a variety of issues as well as to incorporate both facts and the relationships among them as we gather empirical evidence and understand its implications. Instantiations of the conceptual model, in the form of specific mathematical or simulation models, help us focus on what is important for the problem(s) we are working on.

The above example is meant to acquaint the reader with the basics of model building that could be used to explore a *set* of issues rather than modeling that is so specific that one needs to start from scratch if an important aspect of the problem or our understanding of the problem changes. This is what happens when one skips the development of a conceptual model and goes directly to a specific instantiation. All too often, simulation models have built-in (hard-wired) sets of assumptions or represent a partial formulation of the real problem (a limited set of concepts or variables).

Conceptual models represent our current state of understanding and provide a firm foundation to test and improve our understanding. Without a conceptual model to serve as a means to organize what we know, efforts to improve our understanding will be less efficient and less effective.

The next chapter deals with the nature of **Command and Control**. With the understanding of what a conceptual model is and why we want one from this chapter, and a basic understanding of **Command and Control** from the following chapter, the reader will be well-prepared for the step-by-step development of a conceptual model of **Command and Control** that begins in Chapter 5.

CHAPTER 4

COMMAND AND CONTROL

OVERVIEW

Although the purpose of **Command and Control** has remained unchanged since the earliest military forces engaged one another, the way we have thought about **Command and Control** and the means by which the functions of **C2** have been accomplished have changed significantly over the course of history. These changes have resulted from the coevolution of **Command and Control** Approaches with technology, the nature of military operations, the capabilities of forces, and the environments in which militaries operate.

Given that this book is focused on **C2** for transformation, the history of **C2** is not discussed in any detail.[17] In fact, the reader is arguably better off approaching the subject untainted by traditional **Command and Control** concepts, given that transformation calls for disruptive innovation, which translates

[17] Readers interested in the history and traditions of **C2** can find useful resources at the Command and Control Research Program Web site: www.dodccrp.org.

into **Command and Control** Approaches that are not linear extensions of the **C2** concepts that have evolved over time because, for the most part, they represent adaptations to a set of conditions that is no longer applicable.

Functions of Command and Control

Command and Control is not an end in itself, but it is a means toward creating value (e.g., the accomplishment of a mission). Specifically, **Command and Control** is about focusing the efforts of a number of entities (individuals and organizations) and resources, including information, toward the achievement of some task, objective, or goal. How **Command and Control** (or management) is or may have been done in industry and militaries should not be equated with why **Command and Control** (or management) is needed or what functions need to be successfully performed to create value.

Definitions of **C2** are incomplete and potentially worthless unless a means is provided to measure existence (presence) or quality. The U.S. DoD definition of **Command and Control** provides the basis for a test that would indicate its existence (e.g., is there a properly designated commander?). But while this may be observable, it does not provide a good means of knowing how well **Command and Control** is being or has been performed. The official DoD definition provides only one way to assess the quality of **C2** and that is to equate the quality of **C2** to mission accomplishment.[18]

---

[18] **Command and Control**: "The exercise of authority and direction by a properly designated commander over assigned and attached forces in the accomplishment of the mission..." Defense Technical Information Center. DoD Dictionary of Military and Associated Terms. Joint Publication 1-02.

The use of mission success as a measure of the "goodness" of **C2** is problematic because the very definition of the mission is a function of command. Hence, a failure to appropriately define the mission, that is, crafting mission objectives that are unattainable and result in mission failure, is in fact a failure of command. However, while **C2** may be necessary, it is not sufficient to guarantee mission success. That is because the success of a mission is dependent on a great many other factors, including the availability of appropriate means and the capabilities and behaviors of adversaries and others.

Even inspired **Command and Control** may not result in mission success despite a well-crafted mission, while uninspired and even incompetent **Command and Control** may be associated with mission success. Also, setting the bar too low (e.g., not being aggressive or bold enough) would, if one were to use mission success as the measure of the quality of **C2**, result in an inappropriate assessment. Therefore, while mission outcomes should be a factor in the equation, the quality of **C2** should not be deduced solely from mission outcomes.

Rather, the quality of **C2** should be directly measured by examining how well the functions of **C2** have been performed.[19] Included in "**C2** functions" is, of course, the crafting of the mission. Thus, this approach allows us to make a distinction between consequences that are appropriate to consider related to the quality of **C2** and those that are inappropriate to consider.

---

[19] Doing otherwise is a fundamental mistake that has reached epidemic proportions in **C2**-related analysis. The problem is that there is a large number of variables (in addition to how well **Command and Control** is accomplished or how appropriate the **C2** Approach is) that affect mission outcomes.

Before we are in a position to "operationally define"[20] **Command and Control**, we need to understand the functions that are integral to it. The context of **C2** can vary greatly. The nature of the tasks at hand differ widely, ranging from the creation or transformation of an enterprise at the strategic level to employing the enterprise in a major undertaking at the operational level, to the completion of a specific task at the tactical level. As the nature of the task differs, so does the nature of the resources involved. These can range from something that can be accomplished with organic assets to something that requires putting together a large heterogeneous coalition with resources of many types.

In general terms, it is useful to make a distinction between (1) **Command and Control** as applied to an enterprise, that is, creating or transforming an entity or association of entities to make it well-suited to its challenges and to the missions it takes on, and (2) **Command and Control** as applied to a specific undertaking. This dichotomy of enterprise-creation versus enterprise-employment allows us to focus on the essential functions involved and a set of metrics that is appropriate for the nature of the endeavor. In recognizing this distinction, we must make certain that appropriate attention is focused on the interdependencies between these levels of **C2** and on their potentially fractal properties.

The following functions are associated with the **Command and Control** (or management) of a given undertaking:

---

[20] Most definitions are not sufficiently precise to provide guidance on measuring the presence or absence of a thing in question or the degree to which it is or is not. Ackoff and Sasieni make a distinction between definitions and operational definitions. Ackoff, Russel L. and Maurice W. Sasieni. *Fundamentals of Operations Research.* New York, London, Sidney: John Wiley & Sons, Inc. 1968. pp. 390-391.

- Establishing intent (the goal or objective)
- Determining roles, responsibilities, and relationships
- Establishing rules and constraints (schedules, etc.)
- Monitoring and assessing the situation and progress

Some students or practitioners of command may feel that the above list misses some of the most important functions of a commander, specifically those associated with that of an inspirational leader and/or those associated with a "protector" who cares for and nurtures those under his or her charge. To *lead* also carries a connotation of proactive change or movement that can be expressed in any number of synonyms, such as guide, conduct, supervise, or direct.[21]

Clearly, good commanders and managers are, in this sense, leaders as well. Therefore, we need to add the following functions to account for the leadership component of command or management:

- Inspiring, motivating, and engendering trust
- Training and education

In many people's minds, command and leadership come together in one person. In practice, that is not always the case. Theory should be agnostic on this issue. We need to allow for the possibility that these functions could be accomplished by multiple individuals or indeed are accomplished in an emergent fashion. There is also the issue of resources. Integral to **Command and Control** or management are the allocation of existing resources and the search for additional resources. The time horizon that can be considered ranges from the immedi-

[21] Merriam-Webster Online Thesaurus, "lead."

ate and near term to the mid and long term. Therefore, we need to add the following function to the list:

- Provisioning

In military organizations, there has always been a distinction between peacetime and wartime (a distinction that may be losing its relevance). Hence, **Command and Control** concepts must apply to the full range of scenarios from peacetime engagement to high-end conflict. In the following sections, each of the functions identified above are discussed.

Note that the functions of command establish the guidance for and the focus of the functions of control. Hence, every command function is at least partly an instruction to the relevant elements of control.

Establishing intent

Without a "mission" or purpose, **Command and Control** and/or management do not make sense. Their *raison d'être* is to accomplish or achieve something. Although it may, in trivial cases, be enough to simply specify an objective, the crafting of a mission involves far more than this. In addition to the specification of an objective, it is also important to address the risks that are "acceptable" in pursuing the objective. In crafting a mission, the uncertainties that exist as well as an understanding of what is and is not controllable must be factored in. Undertaking infeasible missions is a failure of command.

While the determination of intent and its communication to mission participants are traditionally thought of as responsibilities of the one who is in charge or in command, this does not have to be the case.

In the NATO sense of C3,[22] where consultation is the first C, intent is derived from a process involving multiple parties. There are numerous instances where there is no supreme or higher authority that can, in practice, determine intent. What is important is that the behaviors of the entities involved (individuals, organizations, and systems) act as if they are working toward some common purpose. Thus, intent may or may not be (1) explicitly communicated, (2) consciously or formally accepted, or (3) widely shared.

The DoD definition[23] of commander's intent,

> "a concise expression of the purpose of the operation and the desired end state that serves as the initial impetus for the planning process. It may also include the commander's assessment of the adversary commander's intent and an assessment of where and how much risk is acceptable during the operation,"

personalizes the concept by associating it with a commander, takes an operational (as opposed to strategic or tactical) perspective, and assumes the existence of a formal planning process that precedes execution. More importantly, this defini-

---

[22] According to the NATO glossary, Consultation, Command, and Control (C3) are "the responsibilities and activities of political, military and civil authorities in political consultation, including crisis management, nuclear consultation, and civil emergency planning. The term also applies to the authority, responsibilities and activities of military commanders in the direction and coordination of military forces and in the implementation of orders related to the execution of operations."

[23] Defense Technical Information Center. DoD Dictionary of Military and Associated Terms. Joint Publication 1-02.

tion views the articulation of the nature of the risks that are acceptable as optional. Because intent must include a position on risk to be meaningful, risk can only be implicit if education, training, and doctrine provide sufficient *a priori* understanding and if the situation is "normal."

Early writings about Network Centric Warfare used the term *commander's intent* (from which this definition was subsequently derived). This circumscribed the way that many thought about NCW and has limited its potential. A better instantiation of this idea is *command intent.* Even better would be just the word *intent.* Command intent is a reflection of a collective rather than of an individual. It is a better term because first, it is more reflective of real-world situations (e.g., coalition operations), and second, it opens up the aperture as far as organizational forms of interest. The book *Command Arrangements for Peace Operations* (1995) focuses on the challenges of a reality that does not fit with orthodox views of unity of command, and suggests *unity of purpose* as a more realistic principle than unity of command.[24] Command intent is consistent with unity of purpose without the requirement for a single authority or unity of command. Using only the word *intent* is best because it does not assume the origins of intent and hence allows one to focus on the function of intent and how well intent is established.

Intent is an expression of purpose. As such, the appropriateness of the purpose is a legitimate subject for deliberation. However, for the purposes of this book, the assessment of the quality of intent will be limited to the quality of its expression (the degree to which it is understood) and the quality of its

---

[24] Alberts, David S. and Richard E. Hayes. *Command Arrangements for Peace Operations*. Washington, DC: CCRP Publication Series. 1995. p. 25.

commonality (the degree to which it is accepted). The exception is when an expression of intent is a subset of a higher intent in an echeloned organization. In these cases, an added criterion, the degree to which the intent is consistent with higher intent, needs to be considered.

The expression of intent needs to be examined in the context of the situation and the organization (the entities that comprise the actors). This context fills in the blanks and serves as the filter through which intent is viewed. Understanding within the organization will be influenced by a host of factors including culture, team hardness, and characteristics of individual entities including their experience and behavioral characteristics. These factors will affect how an individual or organization perceives intent and therefore whether or not the expression of intent is sufficiently well-understood to (1) focus actions and (2) achieve a sufficient level of shared awareness. Thus, measures of **C2** quality need to include consideration of the existence of intent, the quality of its expression, the degree to which participants understand and share intent, and in some cases, the congruence of intent.

Determining roles, responsibilities, and relationships

Command and Control implies the existence of more than one individual or entity. In most endeavors, different entities play different roles. The determination of roles, responsibilities, and relationships serves to enable, encourage, and constrain specific types of behavior. The resulting behaviors create patterns of interactions that emerge from a given set of initial[25] conditions or "pre-existing environment." In our con-

---

[25] "Initial" is always relative to some specific point in time.

sideration of new network-centric **C2** Approaches, one important behavior that needs to be understood is collaboration. The nature and extent of the collaborations that will take place will be, to a great degree, determined by the initial conditions. Ultimately, these patterns of behavior are most important, as they will determine the ability of the enterprise to accomplish its missions. It is the function of command to establish these initial conditions, that is, to define and assign these roles and the nature of the interactions that should and should not take place. The approach that will be taken to control is also included in the initial conditions.

Traditionally, the flow of information has been tightly coupled to the command relationships. More recently, information flows have been freed from hierarchical and stove-piped patterns of distribution. Now the roles, responsibilities, and relationships (including the nature of the information-related interactions that should take place) need to be specified separately from the specification of other roles, responsibilities, and relationships. For example, the move to a post-and-smart-pull paradigm involves a departure from traditional military patterns of information flows and information-related roles and responsibilities.

Who, if indeed any specific person or entity, determines the allocation of roles and responsibilities and the nature of the relationships that will exist will differ from situation to situation. Traditional notions of **Command and Control** assume a set of predefined hierarchical relationships that, for the most part, are fixed. But neither the existence of a hierarchy nor the static nature of relationships and assignments need be assumed. Roles, responsibilities, and relationships (or a subset of these) may be self-organized and may change as a function of time and circumstance.[26]

Measures of the quality of the organization (i.e., the ability of a particular arrangement of roles, responsibilities, and relationships and their dynamics to perform the functions needed to accomplish the intended task) should include consideration of (1) the completeness of role allocation (are all necessary roles and responsibilities assigned?), (2) the existence of needed relationships, and (3) whether or not the assignees know and understand what is expected of them (in the satisfaction of their roles). Issues of role overlap and role gaps are also relevant.

Included among the functions that need to be performed is ensuring that intent is known and understood. The determination of what other functions are needed depends on the nature of the situation and the nature of the organization. Some decisions will need to be made and actions taken. The allocation of decision rights and the assignment of responsibilities for action are a reflection of a topology or mapping of roles and responsibilities. While it is conceivable that no interactions will occur between and among entities, this is merely an endpoint on a spectrum and is unlikely in any major undertaking.

The nature of the interactions among entities is, arguably, the critical element in the tenets of Network Centric Warfare and the principles of Power to the Edge. As such, the interactions that are permitted and those that actually take place need to be characterized and observed.

---

[26] For more, see the ONR-sponsored research initiative: Handley, Holly. "Adaptive Architecture for Command and Control." <http://viking.gmu.edu/a2c2/a2c2.htm> April 2005.

Establishing rules and constraints

Rules of behavior and constraints that govern and shape participants' decisions and actions are both "fixed" and "variable." Those that are (relatively) fixed are a reflection or manifestation of human nature and/or those that are expressions of culture (e.g., social, organizational, or professional). Those that are variable pertain to a given situation (e.g., rules of engagement). Constraints can also be fixed or variable. The nature of the rules and constraints that prevail and the explicitness with which rules and constraints are communicated depend on the **Command and Control** Approach being used. As with role determination, how rules and constraints have been or could be established varies. Measures of **C2** quality need to include consideration of the extent to which rules and constraints are understood and accepted. Whether or not the rules and constraints are appropriate or necessary is altogether another matter, and is probably best measured by the relative[27] ability of the enterprise to achieve its purpose, accomplish its missions, or meet its objectives.

Monitoring and assessing the situation and progress

Once intent has been formed (although not necessarily expressed), the clock is running. The time it takes to put the set of initial conditions in place has proven to be a critical determinant of success in a class of missions characterized by a dynamic operating environment. This set of initial conditions is subject to change as the situation changes.

---

[27] *Relative* in this instance implies that a comparison between one enterprise with a given set of rules and constraints and a comparable (in terms of resources) enterprise with a different set is possible.

Thus, an integral part of any **Command and Control** Approach is how changes are recognized and adjustments are made. The ability to recognize a need to change and the ability to adjust are associated with agility. Adjustments may take the form of a change in intent, in the expression of intent, or in the manner in which it is communicated. It may take the form of changes in roles and responsibilities or relationships and it may take the form of changes in rules and constraints.

The DoD definition of **Command and Control** implicitly assumes that plans are developed and then executed. In this case, monitoring and assessing are part of both the planning process and the execution process. In fact, monitoring and assessing define the link between these two often distinct and sequential processes. But there is no need to assume the existence of a planning process or one that is separate from execution (operations). In **Command and Control** Approaches that do not employ formal plans, the functions of monitoring and assessing are accomplished differently.

Measures of **C2** quality therefore need to consider whether changes in circumstances are noted and how quickly they are noted, as well as the appropriateness and timeliness of the response (a change in what "control" variables apply or a change in the "acceptable" range of a control variable).

Inspiring, motivating, and engendering trust

These three interrelated functions, normally associated with leadership, determine the (1) extent to which individual participants are willing to contribute and (2) the nature of the interactions that take place. The effects, the degree to which participants are inspired, motivated, and trusting of each other, and the products and services that are provided potentially affect transactions across the information, cognitive, and social domains. The objects of trust are varied. They include individuals, organizations, and information collectors, as well as equipment and systems. Individuals and organizations will be perceived, and may be stereotyped by role or function. For example, differing degrees of trust may be an initial default depending on whether the relationship is superior-to-subordinate, peer-to-peer, or organization versus organization. These three factors will, for example, affect how participants perceive information provided by others and their willingness to be dependent on others for support. These factors should also be expected to affect the nature of collaborative arrangements.

When the issue is the nature of a mission capability package (MCP) that would work best under a set of conditions, then these parameters need to be considered as part of the given. When thinking about the future of the force one can, by making appropriate investments and making specific policy decisions, affect the values of these parameters (e.g., a volunteer force or a draft, the allocation of responsibilities between active duty and reserve components, and investments in training and education). Investment patterns, in turn, determine the suitability of other components of future mission capability packages (by determining the initial conditions to support alternative **Command and Control** Approaches).

Training and education

The professionalism and competence of a force have a profound effect on the degrees of freedom available when considering ways to organize and alternative **Command and Control** Approaches. Just like the factors considered above, certain levels of training and education are, for all intents and purposes, a given for current operations. They can be modestly affected in the short run, but are capable of being transformed for a future force.

Many **Command and Control** concepts that we have inherited are attempts to deal with a force with given levels of education and/or training.[28] Education is different from training. Education is broader and provides a more thorough understanding of a subject, while training allows individuals to develop and act upon their own expertise. New **Command and Control** Approaches will inevitably require specific education and training (new knowledge and skills).

However, the degree to which a more highly educated or more proficient force than the one we have today is required is an open question. For example, as we began to automate processes and employ more and more sophisticated weapon systems, some feared that we would be building a force that would only work if our soldiers, sailors, airmen, and marines all were degreed engineers. This, of course, is not the way it turned out. Rather, we have a situation today where those entering the force are more comfortable, knowledgeable, and experienced with computers and networks than their more senior colleagues. The reason is that society does not stand still; the Information Age has brought with it changes that have greatly influenced the level of computer literacy and the experiences of our younger generations.

---

[28] The issue of subordinate attributes is addressed in:
Alberts and Hayes, *Command Arrangements for Peace Operations*. p. 74.
Alberts, David S., John J. Garstka, Richard E. Hayes, and David T. Signori. *Understanding Information Age Warfare*. Washington, DC: CCRP Publication Series. 2001. p. 176.

Provisioning

Resources are critical to the success of any endeavor. Provisioning must be thought about from both an enterprise and mission perspective, and from both a short-term and long-term perspective. If one takes a long-term enterprise view, it is about what we might call "shaping the force." This includes the development of mission capability packages[29] and the mix of the capabilities they provide.

Provisioning in a mission context is almost always looking at the immediate and short term and is about allocating available resources and sustaining effort over time. From an enterprise perspective, it is about allocating across missions, and from a mission perspective, it is about allocating resources among participants and over time. Provisioning, at the mission level, includes efforts to obtain more resources from both organic and non-organic sources. Available resources are a critical factor in determining the feasibility of satisfying intent and the appropriateness of organizational arrangements. How well resources are allocated and utilized is often the determining factor in whether or not the intended purpose is achieved. There are many ways to allocate resources among entities and there are many ways resources are matched to tasks. Each of these has the potential to result in different degrees of effectiveness and/or agility. Thus, a measure of the quality of **C2** needs to include a measure of the effectiveness of resource allocation.

---

[29] Alberts and Hayes, *Power to the Edge*. pp. 223-231.
Alberts, David S. *Information Age Transformation*. Washington, DC: CCRP Publication Series. 2002. pp. 73-78.
Alberts, David S. "Mission Capability Packages." National Defence University: Strategic Forum. Institute for National Strategic Studies. Number 14. 1995.

In the mid to near term, resources can be created by making investment choices (e.g., in materiel or training). Thus in the mid to long term, one has a chance to affect the nature of the resources that will be available at some time in the future.

DISTINGUISHING CHARACTERISTICS

There is always a default set of conditions. The exercise of **C2**, which consists of attempting to accomplish its functions in a particular set of ways, seeks to change these defaults.

The following are essential **C2** functions:

- Establishing intent
- Determining roles, responsibilities, and relationships
- Establishing rules and constraints
- Monitoring and assessing the situation and progress
- Inspiring, motivating, and engendering trust
- Training and education
- Provisioning

These functions are associated with mission or enterprise **C2**. They can be accomplished in very different ways. These differences boil down to how authority and relationships are determined, how decision rights are distributed, the nature of the processes involved, how information flows, and the distribution of awareness. Specifying how these functions are to be accomplished determines a particular **Command and Control** Approach. The totality of all possible approaches forms a **C2** space. The dimensionality of this space, as well as the identification of some areas of interest in this space, is the subject of Chapter 6.

The next chapter will provide an overview of a conceptual model of **Command and Control** containing such a space as a set of controllable variables.

Distinguishing characteristics

Chapter 5

C2 Conceptual Model Overview

The conceptual model of **Command and Control** presented in this book has been specifically designed to facilitate exploration of new **Command and Control** Approaches. In formulating this model, we have sought to encourage exploration of alternative **Command and Control** Approaches based on a value chain derived from the tenets of NCW and concepts that adhere to Power to the Edge principles. This exploration requires the ability to make in-depth comparisons between traditional, network-centric, and other approaches that have only been hinted at or may not have been conceived of at this time.

For this reason, the model needs, at a minimum, to explicitly incorporate concepts that are associated with NCW and Power to the Edge, as well as the concepts that underpin traditional Industrial Age **Command and Control** Approaches. The inclusion of these concepts (e.g., shared awareness) should not be taken as an assertion of their importance or significance, but simply as a recognition of the fact that without their inclusion one cannot reasonably explore and assess hypotheses related to the dynamics and value of network-centric **Command and Control** Approaches.

The term **Command and Control** carries with it a host of legacy presumptions and assumptions, and the self-identification of military professionals trained in earlier eras. If one accepts that the term **Command and Control** is inseparable from these cherished traditions and self-identifications, then the model discussed here would not be (for some traditionalists) a **Command and Control** model. This model does indeed facilitate the exploration of collective behavior and its relationship to the degree to which the traditional functions of **Command and Control** are accomplished.

This is an important distinction because it goes to the heart of what we mean by **Command and Control**. On the one hand, there are many who see **C2** as a particular solution; on the other hand there are others who see **C2** as a challenge related to the accomplishment of a set of functions and who place no "traditional" constraints on how these functions are accomplished. The former have constructed models that assume that **C2** is equivalent to current doctrine and processes. The latter seek models that can represent a spectrum of **C2** Approaches. Thus, the model presented here differs from traditional models of **Command and Control** in the nature of or absence of assumptions and constraints that affect interactions among participants in traditional military organizations.

Most of the models in use today incorporate implicitly or explicitly traditional **Command and Control** processes. They therefore incorporate existing organizational structures and concepts of operation. Given that the power of NCW derives from the coevolution of concepts, **C2** Approaches and organizations, information-related capabilities, other technologies, and the knowledge, skills, and expertise of people, models that have built-in organizations, doctrine, and processes will not be able to accurately reflect the impact of network-centric capabilities and any conclusions drawn from these models will be invalid. By developing and employing the kind of model presented here, we can analyze the virtues and shortcomings of non-traditional **Command and Control** Approaches.

Bottom up or top down?

Models can be built from the bottom up or from the top down (such as the approach taken in the example in Chapter 3). The bottom-up approach starts with the identification of a set of variables that are felt to be relevant, while the top-down approach starts with the identification of a minimum set of concepts. Regardless of which approach is initially taken, model development usually involves both, and if the model is successful there will be a convergence. The same test of success can be applied for the results of either approach. This test is whether or not the variables/concepts constitute a set that is both necessary and sufficient for the intended purpose of the model. Two recent serious efforts at developing a comprehensive conceptual model for exploring **Command and Control** took different developmental approaches. The first was a joint ASD-NII/OFT effort[30] that took a top-down approach based on the tenets of NCW. The second was a NATO RTO effort[31] that took a bottom-up approach.

---

[30] Office of Force Transformation. "NCO Conceptual Framework, Version 2." Prepared by Evidence Based Research, Inc. Vienna, VA. 2004.

[31] NATO RTO establishes SAS Panels to advance the understanding of operational analysis and technology. SAS-050 explored a new **Command and Control** model. NATO SAS-050, "SAS-050 Conceptual Model Version 1.0."

Both efforts identified a fairly large number of relevant variables.[32] Although the groups had a few members in common, they were significantly different. The initial lists of variables were compared and found to be equivalent for all intents and purposes, after differences in terminology were taken into consideration. What differed was the way in which each group organized these individual variables into compound variables (functions of two or more variables) and concepts (collections of variables).

As previously discussed, a conceptual model needs to be instantiated before it can explore the behaviors represented. A conceptual model may have a large number of instantiations that differ in the specific variables that are included and the nature of the relationships among these variables. The appropriateness and value of the instantiated model depends on the issues to be explored and the assumptions made. The top-down approach begins with a conceptual model and, during a process of instantiation, identifies specific variables and relationships. The bottom-up approach provides a list of variables that can be organized in a variety of ways. Given that a conceptual model is a reflection of its purpose, it is not surprising that the NATO group, consisting of individuals with different interests and different experiences and which took a bottom-up approach, had great difficulty agreeing on the concepts to be included in a conceptual model. The approach taken here is a top-down one because getting the concepts right will help ensure that the instantiations that we and others develop consider all of the essential pieces of the problem. This approach is also much easier for a reader to follow.

---

[32] Currently, there are 200 variables in the ASD-NII/OFT model and 336 in the NATO model.

Essential Concepts

Figure 4 is a generic version of the conceptual model presented in Chapter 3. This point of departure for the development of a conceptual model is designed to help (1) understand the differences between new network-centric and Power to the Edge **C2** Approaches and more traditional approaches, (2) understand when each type of approach works and does not work, and (3) explore the NCW value chain to guide the design of **C2** Approaches that work well under a variety of conditions, (i.e., to develop agile **Command and Control**). While the conceptual model depicted in Figure 4 can accomplish this if the right set of variables is included in an instantiation of the model, the conceptual model as formulated does not identify the key concepts (variables and associated value metrics) that are needed for our purposes. In this chapter, we will add to and modify this formulation of a **C2** model to better suit our needs.

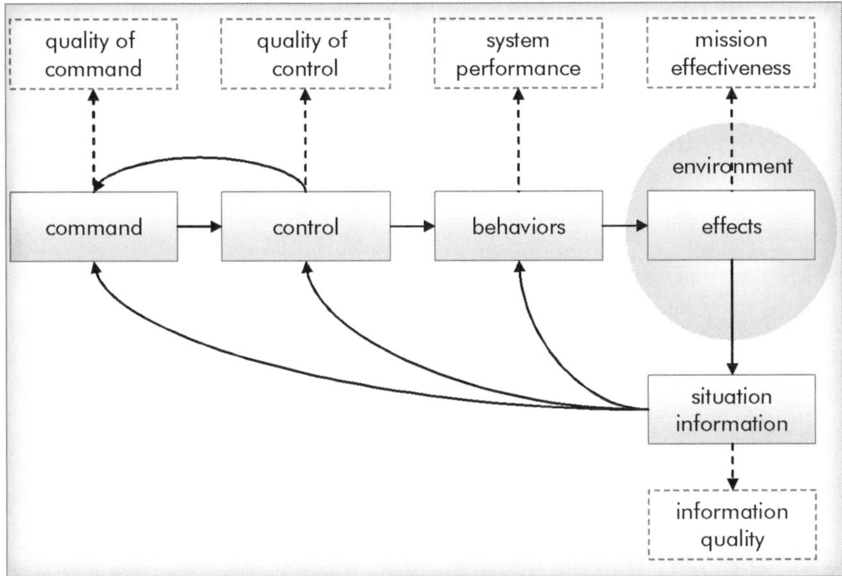

FIGURE 4. C2 CONCEPTUAL MODEL

The **C2** Conceptual Model depicted in Figure 4 is elemental or fractal. An enterprise of the complexity necessary to undertake military and civil-military missions will have many concurrent, nested, and even overlapping instances of this elemental model, each one of (or collection of) which may exhibit different **Command and Control** Approaches. At the enterprise level, the functions associated with command will determine the number and nature of these fractals and the relationships among them. Thus, if we consider Figure 4 to be a view at the enterprise level, then there will be a great many "little Figure 4s" contained in the enterprise view of the behaviors box, or for that matter, the boxes for command and for control. Command at one level determines the conditions under which fractals that are within their purview operate. There will be cases of sovereign fractals in which the fractals are not nested but have peer-to-peer and/or overlapping relationships. In these cases, the functions associated with **Command and Control** are achieved in a manner different from that of traditionally nested fractals.

We begin our examination of this model by looking at the functions and desired states sought by efforts to design or establish effective **Command and Control**.

Command and Control

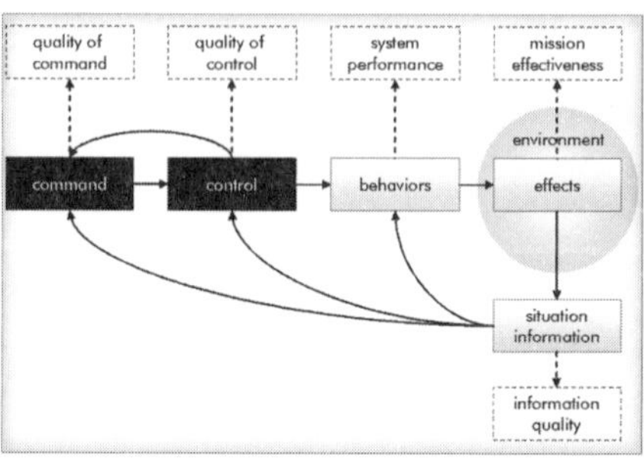

How an enterprise chooses to accomplish the functions associated with **Command and Control** and the impacts and influences associated with the accomplishment of these functions need to be at the heart of a conceptual model of **C2**. Rather than treating **Command and Control** as a single concept, we have chosen to separate command from control to maintain the greatest degree of flexibility. This enables us to examine each concept on its own and combine different approaches to each in ways that have, as far as we know, not been considered before. Thus, we start constructing our conceptual model with two boxes, one representing the concept of command, and the other representing the concept of control. Together, these two boxes define the **C2** space with some points in this space corresponding to traditional **Command and Control** Approaches. Concepts consist of one or more variables. In the case of command and in the case of control, specific combinations of variables serve to define the dimensions that bound the space of possibilities. The values of these variables serve to completely specify a unique approach to accomplishing the functions associated with either command or control. This **C2** Approach space is explored in detail in Chapter 6.

Functions of Command and Control

The **Command and Control** functions discussed herein are applicable not only to military endeavors but also to civil-military and indeed to civilian and industrial enterprises. In the previous chapter, the following **C2** functions were identified:

- Establishing intent
- Determining roles, responsibilities, and relationships
- Establishing rules and constraints
- Monitoring and assessing the situation and progress
- Inspiring, motivating, and engendering trust
- Training and education
- Provisioning

Each of these functions can be seen in the context of a particular time horizon. Provisioning, for example, is constrained to the allocation of existing resources for current operations. For the purposes of the development of this conceptual model, the focus will be on the mission, that is, current operations or the employment of all or part of the enterprise. However, *current* refers to the context in which the functions of **C2** are performed and does not refer to the present time. Thus, for the purposes of accomplishing the functions of **C2**, resources are assumed to be fixed (including established capabilities for resupply and repair). However, because the operations we are thinking about will take place sometime in the future, the nature of the resources that are available are not assumed to be those that are available today. In fact, a major purpose of the model is to support the design of a network-centric mission capability package consisting of coevolved concepts of operation, organization, **Command and Control**, materiel, personnel, education, and training. Presumably, this coevolution will be accelerated by analysis and experimentation driven by this model or other promising approaches.

Prior to the commencement of an operation, intent (a command function) needs to be established (by definition). This intent can consist of merely recognizing that there is a situation that needs to be dealt with or a problem to be solved. It does not require that a solution or an approach be developed. Roles, responsibilities, and relationships may be predetermined or they may be established or modified to suit the circumstances (intent and the situation). The establishment of a role determines whether or not the entity is considered part of the team or part of the environment. Likewise, rules and constraints and resource allocations can be predetermined or tailored to the situation. For the purposes of this discussion (one of current operations), we will assume that the degree to which individuals and organizational entities are inspired, the level of trust they have in one another, their motivations, and their level of training and education are, in fact, initial conditions. That does not mean that we will not consider a learning curve or changes in trust levels during an operation, but that these changes will result from the interactions between initial conditions and the operating environment or circumstances.

Once an operation begins (and this dates from the establishment of intent, not from the commencement of a response, where response can include pre-emptive action), intent can change, as can roles, responsibilities, allocations of resources, and the like. All of these changes to the set of initial conditions, with the exception of a change to intent, should be considered control functions. Changing intent is a command function. The ability to make timely and appropriate changes is directly related to the agility of the specific instantiation of a **C2** Approach. Given the complexity of the 21st century security environment and the missions that 21st century militaries are and will be called upon to accomplish, **C2** agility is perhaps the most important attribute of a **C2** Approach.

Command

The establishment and communication of the initial set of conditions, the continuing assessment of the situation, and changes to intent are the functions of command that we will focus on for this model. Thus the "products" of command, as depicted in Figure 5, will directly determine, impact, influence, or moderate the following:

- Intent
- Allocation of roles and responsibilities
- Constraints on actions
- Awareness of the above, including alternative possible futures
- Nature of the interactions among participants
- Allocation of resources including:
  - Information
  - Personnel
  - Materiel

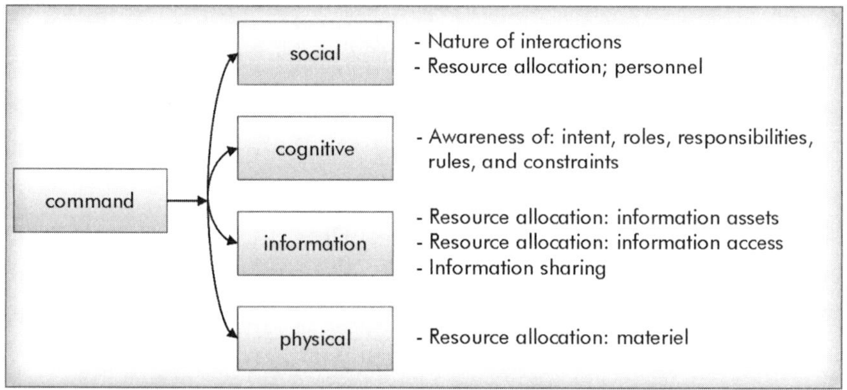

FIGURE 5. OUTPUTS OF COMMAND BY DOMAIN

A priori, command will affect the initial conditions when undertaking a particular mission. These initial conditions include, for example, the resources available for allocation and existing levels of trust, education, and training. Among the resources that will be determined *a priori*, the resources associated with **C2** (i.e., information- and communications-related capabilities) are central to analysis and hence are essential components of a model of **Command and Control**. The variables that define these resources will, for example, directly affect the availability of information and the ability to share information.

The ability to exercise command (the accomplishment of the functions associated with command) is affected or influenced by, among other things, the quality of information available. Therefore, command influences the ability to command over time. Put another way, command sets the conditions under which **C2** is carried out. To oversimplify, command prescribes **Command and Control** processes. Thus, command inherits a set of initial conditions that are a result of previous, longer-term command decisions, and in turn sets the initial conditions for the current operation(s), including shaping the conditions and setting the rules for control.

Control

The function of control is to determine whether current and/or planned efforts are on track. If adjustments are required, the function of control is to make these adjustments if they are within the guidelines established by command. The essence of control is to keep the values of specific elements of the operating environment within the bounds established by command, primarily in the form of intent.

Control can be performed in many different ways. Included in the possible approaches to control are both direct and indirect approaches, as well as approaches that rely on different degrees of granularity or specificity. To be most effective, the approach to control needs to be consistent with the approach to command. The inputs to control consist of the initial conditions set by command, including the approach to be taken and intent for the operation(s) at hand. The outputs of control mirror those of command with one exception: intent. However, a function of control is to interpret and express intent (to the extent that is determined by command). Figure 6 incorporates the concept of control.

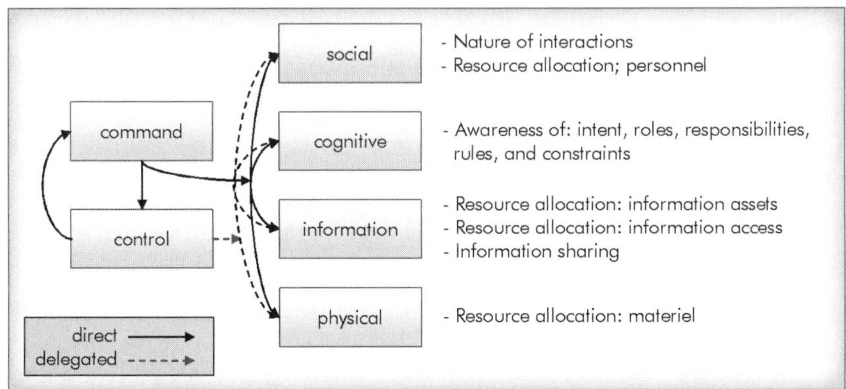

FIGURE 6. COMMAND AND CONTROL AS A FUNCTION OF DOMAIN

C2 sets the conditions and changes the conditions under which information is shared and participants interact, which affects a variety of behaviors.

Behaviors

Behaviors include those actions and interactions among the individuals and organizations that accomplish the functions associated with **Command and Control** (e.g., establishing intent, conveying intent), those that are associated with understanding or making sense of the situation and how to respond, and those that are associated with the response (that is, with creating the desired effects such as maneuver and engagement). The first two sets of behaviors

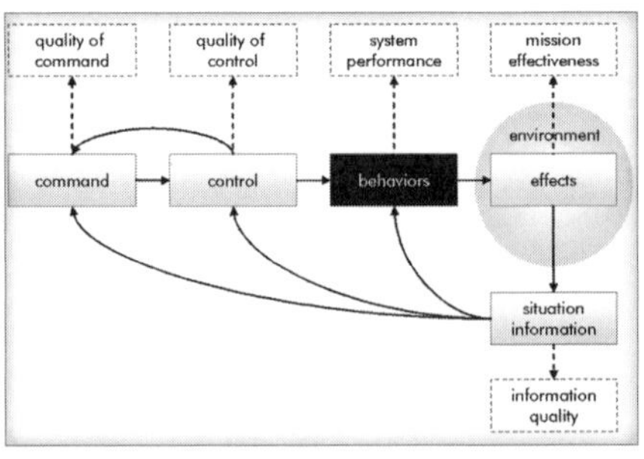

constitute **C2**, the second set of behaviors is a subset of **C2** called sensemaking, and the third set of behaviors can be referred to as actions or execution. All are functions of an enterprise (organization or endeavor).

Traditional treatments of **C2** make the distinction between planning and execution. This has been a particularly poor way of thinking about accomplishing missions because it has led to a separation of these two functions conceptually, organizationally, and temporally. Planning is a part of sensemaking. However, planning activities may or may not result in the production of an explicit plan. Not having an explicit plan is not necessarily bad. It is the process of planning that is important, not the specific plan that is produced.[33] As Von Moltke pointed out, "no plan survives first contact with the enemy."[34] In Industrial Age **C2**, plans include contingencies (alternative sets of missions, asset allocation, boundaries, and schedules to be undertaken when specific sets of recognizable conditions are obtained) that are expressed as sets of "branches and sequels." In today's operations however, characterized by a compression, if not elimination, of meaningful distinctions between strategic, operational, and tactical processes and rapidly shrinking windows of opportunities for effective action, a more fluid approach that allows for simultaneous planning and

---

[33] In the words of General Eisenhower: "In preparing for battle, I have always found that plans are useless, but planning is indispensable."

[34] This quotation is the common version of the writings by Prussian field marshal and chief of staff Helmuth von Moltke (1800-1891). The original quote is: "No operational plan will ever extend with any sort of certainty beyond the first encounter with the hostile main force. Only the layman believes to perceive in the development of any campaign a consistent execution of a preconceived original plan that has been thought out in all its detail and adhered to the very end."
Tsouras, Peter G. *The Greenhill Dictionary of Military Quotations*. London, UK: Greenhill Books. 2004. p. 363.

execution makes more sense. A "plan" becomes the pattern of action in the heads of the participants of the operation that is constantly adjusting to the realities they experience. In other words, the participants are constantly making sense of the situation and taking actions that they believe will synchronize their efforts. The degree to which planning and execution are separate and sequential versus integrated and simultaneous will directly affect **C2** agility and thus force agility in the context of a mission.

Indeed, sophisticated planning is not only reactive (recognizing when the emerging situation deviates meaningfully from the desired or goal states and triggering actions, including contingencies or re-planning, to restore momentum in the desired direction), but also proactive in that it establishes conditions that (1) make future success more likely and/or (2) foresee future developments and undertake activities that will take advantage of them or prevent their negative implications. Indeed, both the agile concept of flexibility (identifying multiple ways to succeed and moving seamlessly between them) and the Soviet goal of reflexive control are manifestations of this more sophisticated class of planning.

C2 Approach

Both the objective of sensemaking and execution and how they are accomplished are determined by **Command and Control**. The approach one takes to **Command and Control** (**C2** Approach) is a function of command. The **C2** Approach selected establishes the conditions under which entities interact. A fuller discussion of the dimensionality of **C2** Approaches can be found in Chapter 6. The **C2** Approach space is depicted in Figure 11. Figure 7 reflects the changes to the conceptual model needed to incorporate the concept of a **C2** Approach, which affects the processes within the shaded area.

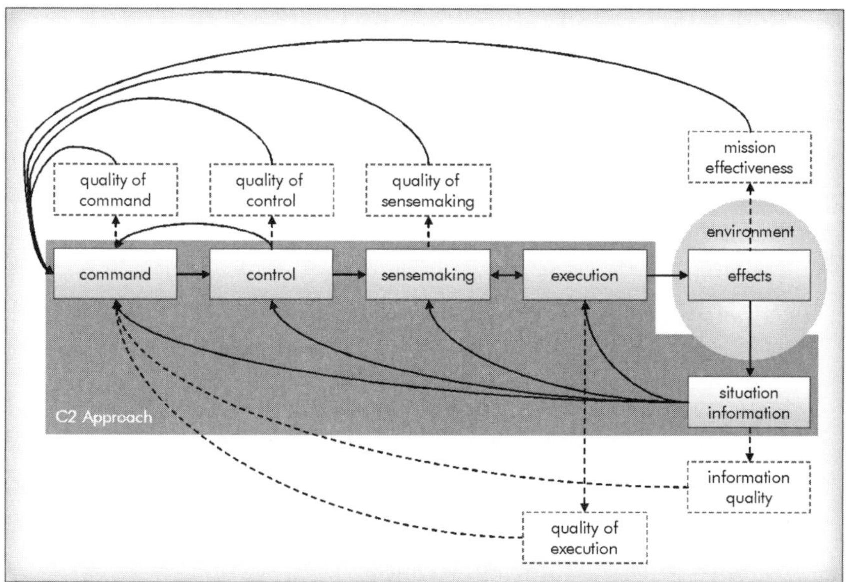

FIGURE 7. C2 CONCEPTUAL MODEL: C2 APPROACH

SENSEMAKING

Sensemaking consists of a set of activities or processes in the cognitive and social domains that begins on the edge of the information domain with the perception of available information and ends prior to taking action(s) that are meant to create effects in any or all of the domains (for example: the employment of kinetic weapons with direct effects in the physical domain and indirect effects in the other domains; the employment of psychological or information operations designed to create direct effects in the cognitive and information domains with indirect effects in the physical domain).

Understanding sensemaking in a network-centric environment, as explained in *Information Age Transformation*, requires an

> understanding of individual and collective processes by which tacit knowledge (e.g., experience, expertise, and culture) is combined with real-time information to identify, form, and articulate appropriate points in an ongoing military operation. These processes can be described in terms of four general capabilities involved in the transformation of real-time battlespace information into appropriate decision events and command intents:
>
> 1. Shared Situation Awareness: the capability to extract meaningful activities and patterns from the battlespace picture and to share this awareness across the network with appropriate participants.
> 2. Congruent Understanding and Prediction: the capability to temporally project these activities and patterns into alternative futures so as to identify emerging opportunities and threats.
> 3. Effective Decisionmaking: the capability to form focused and timely decisions that proactively and accurately respond to these emerging opportunities and threats with available means and capabilities.
> 4. Clear and Consistent Command Intent: the capability to articulate decisions in terms of desired goals/effects, constraints, and priorities that are functionally aligned across the network and with other participating organizations.[35]

[35] Alberts, *Information Age Transformation*. pp. 137-138.

The exercise of command includes the specification and resourcing of sensemaking. Command establishes the conditions under which and the processes by which sensemaking is accomplished. Figure 8 depicts the concepts associated with this aspect of the conceptual model.

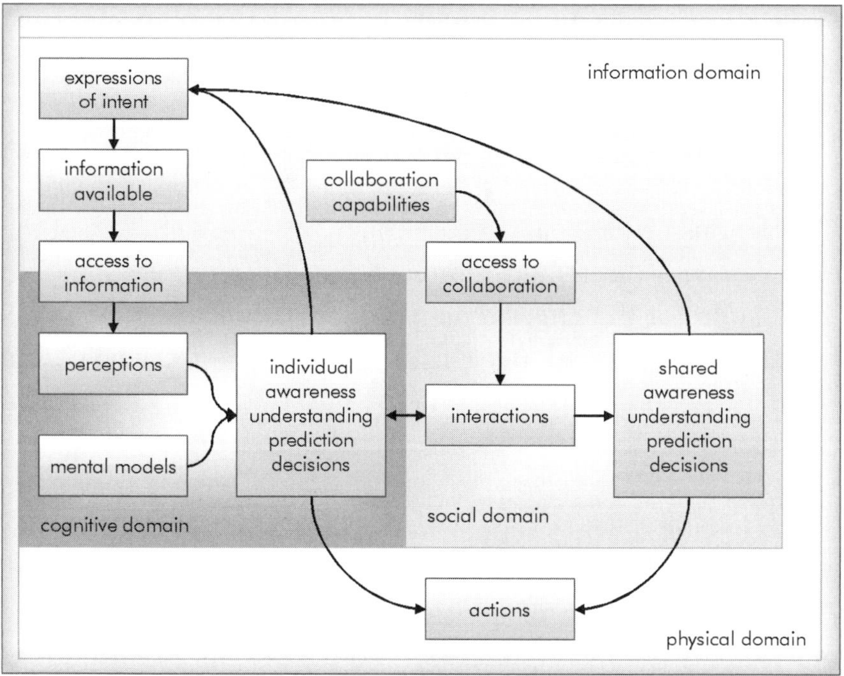

FIGURE 8. SENSEMAKING

In the definition of sensemaking provided here, the cognitive and social processes that are involved in the exercise of **C2** are not included. These are the cognitive and social processes that determine the conditions under which sensemaking takes place and are left as parts of the concepts of command and of control. One might reasonably argue that these cognitive and social processes are indeed about making sense of the situation and shaping a response or refer to these activities as meta-sensemaking—sensemaking about sensemaking.

In the final analysis, it does not really matter what one calls them as long as they are explicitly incorporated into the model. This model separates meta-sensemaking[36] from sensemaking primarily to make sure that it is recognized that an important function of command is to shape sensemaking. This will help to avoid potential confusion when the model is applied to an Edge organization that is capable of dynamically morphing its **C2** Approach.

Execution[37]

The actions involved in execution may take place in any of the domains with direct and indirect effects in multiple domains. The nature of the effects created by a particular action are a function of (1) the action itself, (2) when and under what conditions the action is taken, (3) the quality of the execution, and (4) other related actions.

The selection of what actions to take and when to take them is part of the sensemaking process. This selection is normally the result of a collection of decisions, not an individual one, and may or may not involve collaboration. Thus, the contributions to value from (1), (2), and (4) are associated with the sensemaking process. Of course, the extent to which the execution process is coupled with decisionmaking depends upon the **C2** Approach selected, particularly who interacts with whom and the distribution of information.

---

[36] *Meta* is used with the name of a discipline (in this case, sensemaking) to designate a new but related discipline designed to deal critically with the original one.
[37] For a discussion of execution (means and ends) in a real-world context, see: Prins, Gwyn. *The Heart of War: On Power, Conflict, and Obligation in the 21st Century*. London, UK: Routledge. 2002. Chapter 7.

The purpose of **Command and Control** is to bring all available information and all available assets to bear. Mission success may, in fact, not be achieved even if the best **C2** Approach for the situation is employed and **C2** is executed perfectly. In these cases, other factors dominate. These might include a lack of appropriate means. Thus we, in our explorations of **C2**, are interested not in mission success but in the appropriateness of the **C2** Approach selected and how well it is executed. The effect of **C2** on action effectiveness is part of this consideration. Of course, the operational community is always concerned with mission success. Our perspective is ensuring that **C2** contributes as effectively as possible to that mission success, but our focus must remain on the **C2** arena.

The quality of execution is significantly affected by both the type of **C2** Approach and how well it is implemented. For example, once a target is selected, the ability of a weapon to service the selected target is a function of the quality of the information used to aim the weapon. The quality of a particular unit's ground maneuver (e.g., time to get into position) is a function of the quality of its awareness (individual and shared). This is because units can move faster if they can be assured that there are no enemy units that can either see or engage them or if they can pick a good route and avoid problematic routes. This can be thought of as acting with greater "boldness," but really takes the form of acting with greater confidence.

When empowered by high quality shared awareness, professional forces will often exploit their awareness to seize and hold the initiative. Thus, the conceptual model will need to incorporate the concept of quality of action execution and relate it to the quality of awareness and shared awareness.

The degree to which the actions taken are synchronized is of primary interest because specific actions can become both more efficient and more effective if they are taken together or are properly sequenced. The **C2** Approach that is taken directly affects the degree of action/effect synchronization that can be achieved. Thus, the conceptual model also needs to explicitly incorporate the concept of action synchronization and relate it to the **C2** Approach and the quality of awareness.

PROCESS AND VALUE VIEWS

The conceptual model consists of two kinds of concepts: functional or process concepts and concepts related to value. A generic process view of the conceptual model is depicted in Figure 9.

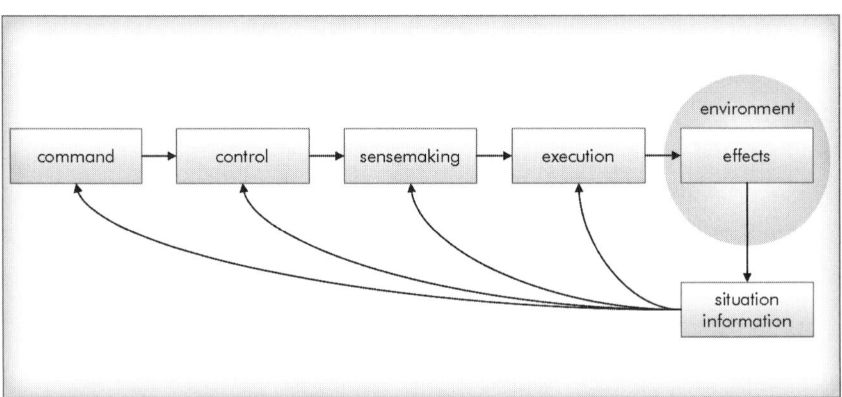

FIGURE 9. C2 CONCEPTUAL MODEL: PROCESS VIEW

For each process or functional concept there is a corresponding value concept. There are several different types of relationships that are of interest in this conceptual model. The first type of relationship of interest is the relationship between or among the process concepts, as depicted by the solid lines in

Figure 4, Figure 7, and Figure 9. The second type involves the relationships between process or functional concepts and related concepts of value. These are shown as dotted lines in Figure 4 and Figure 7.

The third type of interest consists of the relationships that exist between and among the value concepts. Taken together, these relationships define a value chain.[38] Such a value chain is depicted in Figure 10.

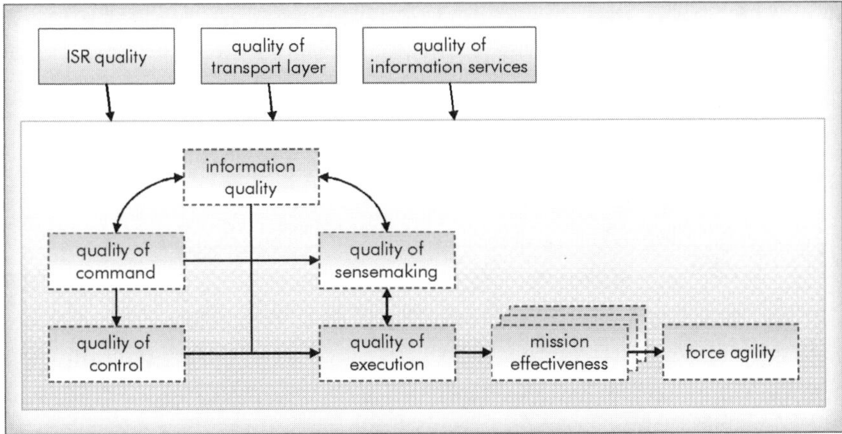

FIGURE 10. C2 CONCEPTUAL MODEL: VALUE VIEW

This value chain incorporates the six quality measures depicted in Figure 7 that correspond to process or functional concepts (e.g., quality of sensemaking). Mission effectiveness is highly scenario-dependent. For the same reasons that we extend our model beyond mission effectiveness to force agility, DoD and other military institutions have moved from a threat-based to a capabilities-based planning paradigm. This helps to move the focus from mission effectiveness in the context of a

[38] A value chain may not be simple, with each link only connected to two other nodes. A value chain may involve branches and loops, as in the **C2** Value Chain.

selected subset of missions to a notion of force agility.[39] Force agility has not simply been added as a measure of value; rather, force agility is being suggested as *the* measure of value of choice. As shown here, it is a function of a number of the other value concepts. The figure also makes a distinction between **Command and Control** as a functional concept and the systems and capabilities that support it. Figure 10 also introduces an illustrative set of value metrics that reflect the quality of the systems that support **C2**.[40] ISR quality, the quality of the transport layer, and the quality of information services reflect our ability to collect, disseminate, and process data, and to interact in the information domain. Included are capabilities that help us visualize and collaborate. These capabilities, in fact, facilitate or constrain selected **C2** behaviors and enable or limit selected **C2** Approaches.

Concepts to Variables

The process and value views presented in the two figures above need to be instantiated before they can be used for in-depth exploration of the range of issues of interest. Instantiating these concepts begins by identifying variables and the relationships among them that are relevant to the issue(s) at hand, and then by identifying additional variables that may have significant effects on the values that these variables take on and/or the nature of the relationships among them.

---

[39] For a discussion of *agility* and its dimensions, see: Alberts and Hayes, *Power to the Edge.* pp. 123-163.

[40] The term *C4ISR* (command, control, communications, intelligence, surveillance, and reconnaissance) is frequently used to refer to the myriad systems that support **Command and Control**.

In the following chapters, we explore both the value and process views of a conceptual model designed to explore a variety of **Command and Control** Approaches. We begin by looking at what makes one approach to accomplishing the functions associated with **C2** different from another. This will result in the identification of a **C2** space. The relevant points and regions within this space will correspond to a set of values taken on by variables that collectively define the dimensionality of a **C2** Approach. These are our primary controllable variables. The choices of the values of these variables will determine the particular **C2** Approach(es) to be examined.

After defining the **C2** space—the space of possibilities that can be explored—we turn our attention to the challenges associated with accomplishing the functions associated with command and with control. These challenges will point us to a set of variables that must be included in the conceptual model because they prescribe the conditions under which **Command and Control** is exercised or they represent desired capabilities or outcomes that we seek by exercising **C2**. The conditions under which **C2** is to be exercised may, under certain circumstances, be "controllable," although these variables are usually thought of as uncontrollable independent variables. Two important instances where they are controllable are (1) during experiments, exercises, or analyses when the values of these variables may be determined, and (2) when one has a choice regarding if and when to conduct an operation when the decision is predicated on a set of acceptable values for these variables. In these situations, we say that "we engage the enemy at a time and place of our choosing." Of course, we may also seek to shape the operating environment in ways that favor our force and improve our likelihood of success.

Concepts to variables

With these two critical sets of variables identified, we will incorporate them into both a value and a process view and then take a look at other variables that we believe to have a significant influence on the values that these variables take on or the nature of the relationships among them.

Concepts to variables

CHAPTER 6

C2 APPROACHES

INTRODUCTION

Perhaps the most pernicious issues in understanding **C2** are (1) the lack of a useful analytical definition and (2) the implicit assumption that **C2** *is* how traditional military organizations perform the functions of **Command and Control**. The definitions used by the U.S. Department of Defense[41] and NATO[42] may be valuable in the legal, institutional, and operational settings for which they were developed, but they are not useful for analysis or research. We and others have previously pointed out the problems with relying on these

---

[41] Again, **Command and Control**: "The exercise of authority and direction by a properly designated commander over assigned and attached forces in the accomplishment of the mission..." DoD Dictionary of Military and Associated Terms.
[42] NATO glossary: **C2** is "the functions of commanders, staffs, and other **Command and Control** bodies in maintaining the combat readiness of their forces, preparing operations, and directing troops in the performance of their tasks. The concept embraces the continuous acquisition, fusion, review, representation, analysis and assessment of information on the situation; issuing the commander's plan; tasking of forces; operational planning; organizing and maintaining cooperation by all forces and all forms of support..."

"institutional" definitions when engaged in research or looking at the future of **Command and Control** functions.[43] This is because these institutional definitions are most often the products of a committee or a coordination process and as such are politically rather than scientifically correct. They also almost always reflect current thinking and processes. These problems include the obvious fact that **Command** and **Control** are different—but closely linked—functions, and that the existing definitions focus on the legal distribution of authority, as well as the failure of these definitions to highlight the unique contributions that we associate with leaders and commanders.[44]

Working with a group of senior professionals from NATO and other nations,[45] we have recently concluded that there is a better approach to defining **Command and Control**. The argument is that there are three key factors that define the essence of **C2** and two important ways that those factors vary within the structures and processes of a given enterprise (service, nation, coalition, or force).[46] These three key factors, which can be thought of as the dimensions of a **C2** Approach, are the:

---

[43] Alberts and Hayes, *Command Arrangements for Peace Operations*. pp. 10-17.
Alberts and Hayes, *Power to the Edge.* Chapter 2.
Pigeau, Ross and Carol McCann. "Re-conceptualizing Command and Control." *Canadian Military Journal*. Vol 3, No. 1. Spring 2002.

[44] We use the term *commander* when referring to the person in a military organization who is assigned that role and the term *leader* when referring to the individual who plays the same role in a non-military organization. Commanders and leaders are individuals who play important roles in managing organizations of all types.

[45] The SAS-050 (studies, analysis, and simulation) was chartered by NATO RTO, as well as Australia and Sweden, to explore and develop a new conceptual model of **Command and Control**. The final report is available at www.dodccrp.org.

[46] A "force" may include interagency partners, international organizations, private industry, and private voluntary organizations.

- allocation of decision rights;
- patterns of interaction among the actors; and
- distribution of information.

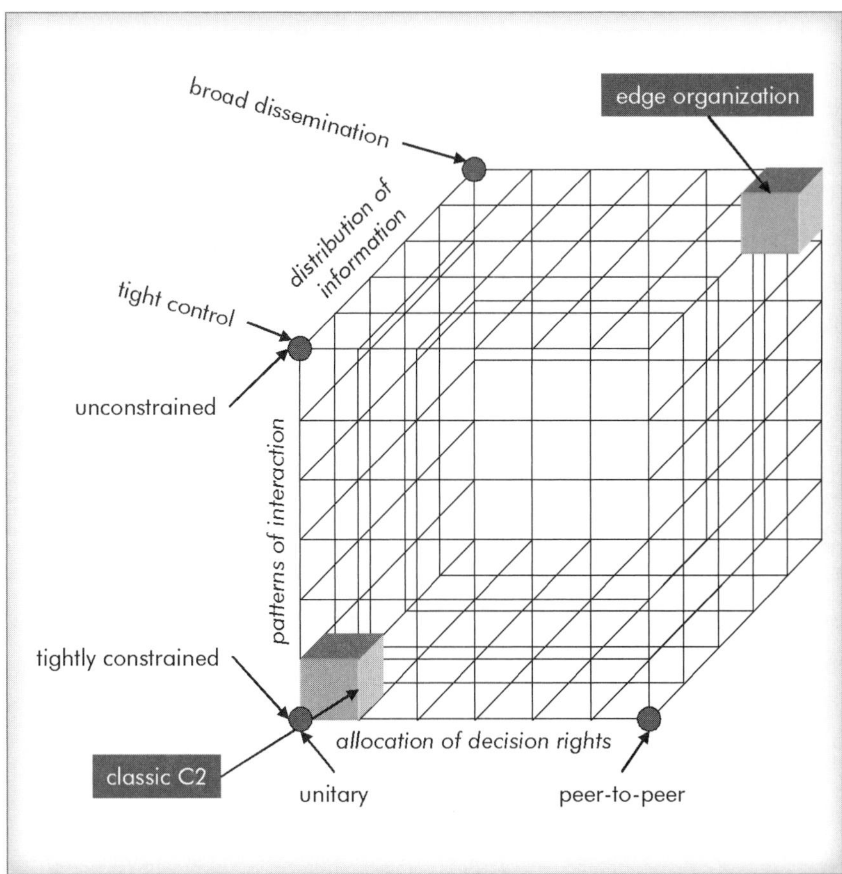

FIGURE 11. THE C2 APPROACH SPACE

We are interested in the *actual* place or region in this space where an organization operates, not where they think they are or where they formally place themselves. We are all familiar with the informal organizations that exist, often operating in very different ways than the official doctrine would dictate. For example, the formal allocation of decision rights may not correspond to the effective (actual or in fact) distribution because of a variety of factors, including traditions, culture, or level of training within a force. Similarly, the formal patterns of interaction or distribution of information desired by the leadership may differ substantially from the way information flows and is distributed in the real world because of informal relationships, linkages, sources, and the real-time requirements of a situation.

We must also allow for the fact that enterprises are not homogenous in any of these three dimensions. In fact, an organization's location in the **C2** Approach space usually ranges across both function and time. For example, with respect to function, intelligence operations may operate in a different part of the **C2** Approach space from logistics, and the conduct of a humanitarian assistance operation may operate with a **C2** Approach that is quite different from the **C2** Approach used for combat operations. Similarly, with respect to time, the crisis management phase of an operation may operate in a way that would not be appropriate if a war broke out. Hence, the **C2** Approach of a given service, nation, coalition, or force may well be best understood as a region or collection of regions within the three-dimensional space rather than, as it is usually thought of, as a point within that space.

Problem space

Clearly, some types of **C2** Approaches will be better suited for certain types of problems. Mapping **C2** Approaches to the types of problems for which they are relatively well-suited requires an understanding of the important ways that problems differ. We posit a problem space defined by the following three dimensions (which may not be orthogonal or totally independent):

Problem space

- rate of change (static versus dynamic)
- degree of familiarity (known versus unknown)
- strength of information position (informed versus uninformed)

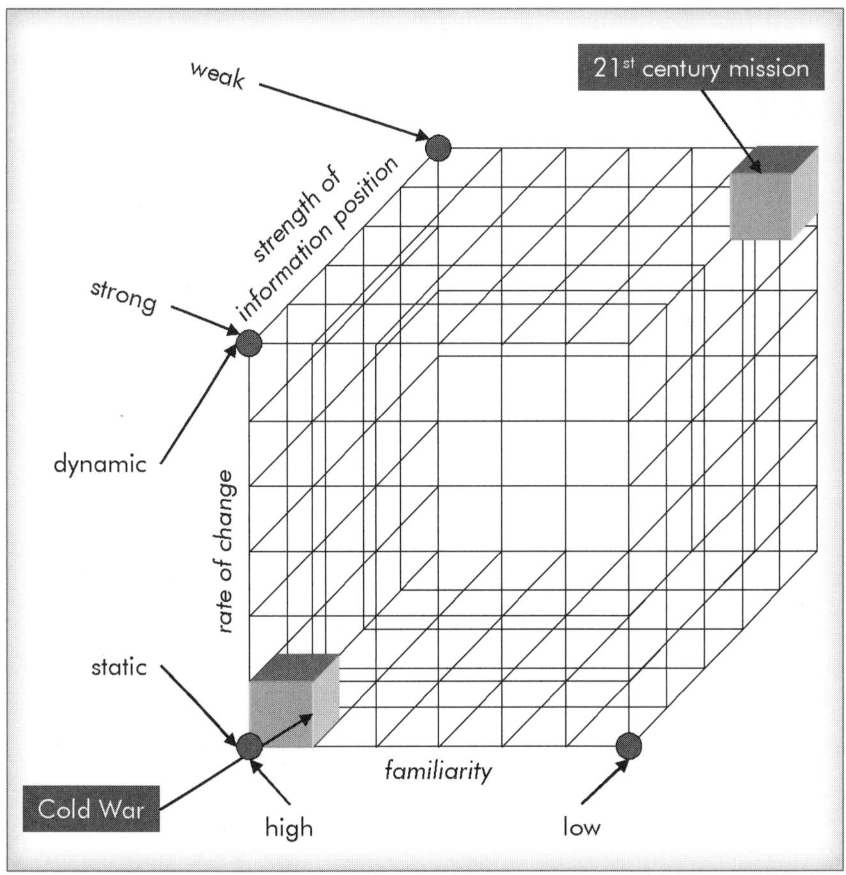

FIGURE 12. THE C2 PROBLEM SPACE

Rate of change

Static problems are those for which the situation itself does not change rapidly. For example, the trench warfare that evolved in Europe during World War I proved to be static—

the front lines, the political, social, and economic operating environments, and the methods of warfare employed changed only slowly. Dynamic problems involve rapid change across all of these features—the location of critical times and places during the struggle change quickly, the operating environment is unstable, and the parties to the conflict innovate frequently and rapidly. Clearly, more static problems are amenable to more centralized decisionmaking in which efforts can be optimized, allow preplanning of the flow of information, and can be controlled by regularized patterns of interaction that involve known groups of specialists. However, in dynamic situations, these classic Industrial Age practices will become impediments to successful **Command and Control**.

Degree of familiarity

When the nature of the problem is well-known (and familiar), everything is relatively simple. First, the information requirements and who needs what information are reasonably well-understood and can be supported efficiently. Second, the patterns of interactions needed are clear. This enables clear decisions about the appropriate allocation of decision rights within the enterprise. In other words, Industrial Age **C2** Approaches can be used efficiently and effectively.

Note that familiarity with a situation is not necessarily a correlate of the degree of situational dynamism present. For example, both NATO and Warsaw Pact military contingency planning for World War III in Europe foresaw a very dynamic battlespace. However, after decades of planning, preparation, and intelligence collection on both sides, each side believed they had a good understanding of how such a conflict would unfold. As a consequence, they developed highly specialized forces and detailed plans for the conduct of this "familiar" conflict.

Of course, confidence that a situation is well-understood does not always translate into effective force development and planning. The French believed they understood the dynamics of an attack from Germany in the later 1930s and committed themselves to a strong linear defense, epitomized by the Maginot Line. However, they found themselves in a very unfamiliar conflict when the Germans unleashed their Blitzkrieg. A more knowledgeable organization, one in which the situation is familiar to a large number of individuals, can distribute decision rights further than one in which less knowledge is present or knowledge is concentrated.

Strength of information position

Finally, situations can also differ in terms of the extent to which the decisionmaking is informed or uninformed. Regardless of the degree of dynamism (though very possibly influenced by that factor) and the degree of knowledge available about it (though, again, very possibly influenced by it), the strength of the information position has an important impact on the applicability of a particular **C2** Approach. As is explained in *Understanding Information Age Warfare*,[47] the information position of an organization is the degree to which it is able to fulfill its information requirements. Hence, a force with very simple information requirements (for example, a terrorist organization) may have a strong information position although it possesses relatively little information. At the same time, a coalition seeking to employ sophisticated weapons and con-

[47] Alberts et al., *Understanding Information Age Warfare*. p. 106.

duct successful counter-insurgency operations may have a great deal of information, but still have a relatively weak information position because of the massive amount of information it requires. At the same time, the extent to which the information available is of high quality (correct, current, accurate, precise enough to support its use, etc.) also influences the information position of a force or an organization. Clearly, a well-informed force can distribute decision rights differently than a weakly informed one. Moreover, a well-informed force should distribute its information more broadly and encourage timely collaboration about what that information means and how to act on it successfully.

All this having been said, identifying the crucial elements of the problem space and matching regions in this space to regions in the **C2** Approach is a high priority. Research should include historical analyses, review of lessons learned in recent and ongoing conflicts, and experimentation. We know, for example, that the level of training and education required for some **C2** Approaches are greater than those required for others. However, we do not know whether current constructs of knowledge and information contained in the NATO **C2** Conceptual Reference Model are adequate to capture the essence of these issues. We also know, for example, that the ability to shift between **C2** Approaches is a method for dealing with dynamic operating environments, but we do not know (1) how great a range of **C2** Approaches a force or organization can muster or (2) whether there are alternatives such as hybrid **C2** Approaches that might make it possible to operate across some relatively broad range of situations.

The discussion that follows starts by examining specific regions within the **C2** space in order to keep the material easier to understand. However, later discussions evolve to deal with a range on each dimension and the regions they define. In addition, we begin with a general appreciation of **C2**, and then move to independent analysis of the command approach and the control approach. While the distinction between these two is critical for analytical purposes, they really must be understood as parts of a unified whole. While military operations can be discussed in terms of their functions of intelligence, operations, logistics, and other functional areas, they ultimately cannot be understood unless the interactions and interdependencies among them are properly appreciated. Similarly, the concept of a **C2** Approach can only be understood in terms of its elements if these discussions are ultimately brought together. One of the valuable results of such an analysis is the ability to assess the compatibility (or lack thereof) between a particular approach to command and a particular approach to control.

In practice, the three key dimensions across which **C2** Approaches differ are not really independent, so showing them as three axes of a cube is something of a distortion. The relationships between them have been illustrated in Figure 13 as a "waterfall chart" because the most fundamental dimension is *allocation of decision rights*, which impacts the other two and, together with *patterns of interaction*, goes a long way toward determining the *distribution of information*. Moreover, the resulting *distribution of decisions* (the "real" distribution that emerges within the dynamics of a situation) may result in a change in the basic allocation of decision rights. This is an example of the type of adaptation (change in work processes and/or organizations) that occurs as part of the agility needed for effective military operations. For example, when a military situation becomes urgent (e.g., an ambush at the tactical level, the realization that an adversary has executed an effective deception plan at the operational level and therefore friendly forces are incorrectly positioned), commanders at lower levels will not (under the doctrine of most modern forces) consult with higher headquarters about deviating from the plan or wait for a new plan, but rather take the initiative by making decisions about how their forces will immediately react. They then inform higher headquarters of what has occurred and the actions they are taking and request support so that they can deal with the ongoing challenge. (Of course, in an ideal world, the other parts of the force, including their higher headquarters, would be able to monitor the situation and would know that they had begun to take initiatives.) If these actions take them outside the existing plans or guidance, they will have altered the distribution of decision rights.

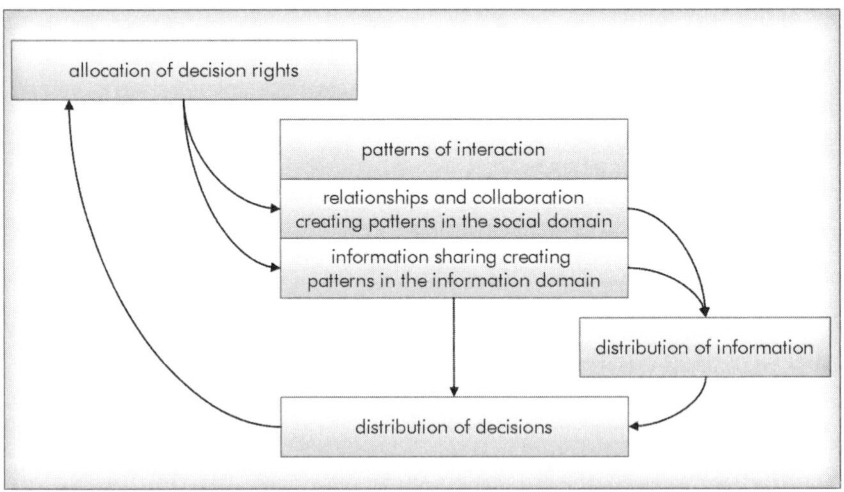

FIGURE 13. THREE KEY DIMENSIONS OF A C2 APPROACH

As is discussed in more detail below, the allocation of decision rights either establishes or enables the establishment of the mechanisms by which those within the enterprise share information and collaborate. These structures, in turn, are major factors in determining the patterns of interaction within the social domain and the information domain. Those patterns, played out over time, have an important impact on the "real" distribution of information within the enterprise.

ALLOCATION OF DECISION RIGHTS

Decisions are choices among alternatives.[48] Decision rights belong to the individuals or organizations accepted (whether by law, regulation, practice, role, merit, or force of personality) as authoritative sources on the choices related to a particular topic under some specific set of circumstances or conditions. The allocation of decision rights is their distribution within the international community, a society, an enterprise, or an organization. In this context, the organization of interest is a military, a coalition, an interagency effort, or an international effort including military elements. There can be different distributions of those rights across functions, echelons, time, or circumstances.

In theory, the allocation of decision rights is a linear dimension with two logical endpoints. At one end of the spectrum is total centralization (all the rights held by a single actor). The totalitarian regimes of Stalin and Hitler were understood to take this form, though it is unlikely that they had the ability to personally control all aspects of their societies on a continuing basis. At the other end of the logical spectrum is total decentralization (every entity having equal rights in every decision, or a uniform distribution). While both total concentration or

[48] This is true even when the decision is to do nothing or when the decisionmaker is not aware of the decision, as in Recognition Primed Decisionmaking (RPD).

centralization and perfect equality or complete decentralization are sometimes found in very small groups, neither of these extremes is likely to be found in any enterprise that would be important to our analyses. Even the ideal "real world" political democracy uses representatives elected by individuals (who possess equal voting rights) in order to make authoritative decisions.[49] Hence, for the purposes of understanding the allocation of decision rights in military organizations or civil-military enterprises, the interesting range of values lies well inside the two extreme points of the spectrum.

Allocation of decision rights in the Industrial Age

The allocation of decision rights in Industrial Age militaries reflects the basic management and knowledge structuring principles of that era. First, the problem of military decisionmaking is decomposed into specialized functional roles using the "Napoleonic"[50] system of command, intelligence, operations, logistics, plans, and so forth. Second, a system of echelons is employed to ensure both appropriate span of control (that is, middle managers that are available to translate guidance from above into actionable directives, to monitor and report activities in the operating environment), as well as what happens to and within the elements of the force, and to act as control agents to ensure that guidance is understood and followed.

Hence, decision rights in Industrial Age militaries are centralized in those responsible for the command function, but also allocated to functional specialists (logisticians, for example) who

---

[49] Even strong representative democracies often use practices (like the U.S. Electoral College or districts of uneven size) that technically violate the idea of precisely equal rights for every individual, usually to protect minority rights.
[50] Alberts et al., *Understanding Information Age Warfare*. p. 192.

work in echelons provided that the choices they make are consistent with the overall guidance provided by commanders. One reflection of the centralization of these militaries is the practice of issuing orders in the name of the commanding officer, even when they may deal with some specialized functional area. Staff members and subordinate commanders have a serious responsibility for ensuring that they understand the guidance from above, which is one reason that a great deal of stress is placed on the Industrial Age concept of *commander's intent* (as opposed to the Information Age variant *command intent*).

We should note that not all Industrial Age militaries have (or have had) highly centralized decision rights. In earlier work, we demonstrated at least six successful 20th century military **C2** Approaches located at some distance from the endpoints of the centralized-decentralized spectrum.[51] Some of these have or currently employ *mission type* orders that devolve (or explicitly delegate) meaningful choices to lower echelons. However, none of the doctrines examined in those analyses go far enough to qualify as either self-synchronizing or Edge approaches.

Allocation of decision rights in the Information Age

Information Age organizations, including militaries, are expected to have minimally centralized distributions of decision rights. An extreme case can be imagined in which there are only emergent distributions and no formal or rule-based distributions of these rights. The tenets of Network Centric Warfare indicate that self-synchronization will be enabled when there is a "critical mass" of shared awareness (in the

---

[51] Alberts and Hayes, *Command Arrangements for Peace Operations*. pp. 67-72. Alberts and Hayes, *Power to the Edge*. pp. 18-26.

presence of suitable doctrine).[52] The success of NCW and Edge C2 Approaches depend on a broader distribution of information and different patterns of interaction (the other key dimensions of a C2 Approach), but they also assume that individual entities have the capacity, information, and means to make effective decisions. Edge organizations[53] not only assume a widespread sharing of information, but (unlike NCW[54]) also require broad distribution of decision rights. Hence, Information Age militaries will require not only better information, better mechanisms for sharing information and collaboration, more knowledgeable personnel, and better trained personnel than their Industrial Age counterparts, but also a different C2 Approach.

Leadership will also need to be different in Information Age militaries. First, it is the responsibility of leadership to ensure the competence of the elements of the force before it is deployed. At the level of creating and maintaining the enterprise, the leadership role or function includes recruiting, equipping, and training a force that is competent to perform the missions it is assigned. General Marshall was credited with playing this role successfully for the United States during World War II.[55] Second, when the enterprise is to be employed, the leadership function includes making the mission

---

[52] See: Alberts and Hayes, *Power to the Edge.* p. 27.

[53] For more, see: Alberts and Hayes, *Power to the Edge.* p. 181.

[54] In NCW, the allocation of decision rights is not directly addressed, while it is in Edge organizations.

[55] After the bombing of Pearl Harbor, General Marshall implemented his plans in order to increase the 200,000 man force to an impressive 8 million. This, of course, required an extensive degree of leadership in order to recruit, train, and equip the larger force. To his credit, this was accomplished without compromising the military's level of competency. For more, see: Cray, Ed. *General of the Army, George C. Marshall: Soldier and Statesman.* New York, NY: Cooper Square Press. 2000.

and broad course of action or approach to accomplishing that mission (together in the form of intent) clear to all those involved, as well as ensuring that they are highly motivated. Third, when employing the force, those with leadership roles must also recognize that not all of the leaders involved in a mission will be in the same hierarchical chain of command, nor will their chains of command necessarily come together at a single point.

Whether the focus is on a military coalition (in which elements of the force are ultimately controlled by the sovereign nations that deploy them) or a civil-military force that includes non-governmental organizations and/or international organizations, there may ultimately be no specific leader with authority over all of the elements. Moreover, while coalition forces often appear to be organized into hierarchies, the forces from each nation often have somewhat different goals, rules of engagement, or constraints imposed by their national leaderships. In a similar way, different agencies in an interagency effort may pursue somewhat different agendas within the overall national guidance or may have differing legal or budgetary constraints than the military. More obviously, the charters and agendas of international organizations, private voluntary organizations, or host governments may not always be fully consistent with those of the U.S. military.

Hence, more than a decade ago we began to discuss command arrangements[56] to emphasize the fact that decision rights may be distributed quite differently in 21st century operations. We have also coined the term *command intent* to replace the singular (and we think incorrect) term *commander's intent*, because there is

---

[56] Alberts and Hayes, *Command Arrangements for Peace Operations*. 1995.

no longer (if there ever was except in specific tactical situations) a single commander present in any reasonably large mission space. Rather, there are sets of leaders and commanders distributed across the functions, organizations, and echelons involved.

A critical challenge for Information Age enterprises including militaries (e.g., organizations that adopt Information Age C2 Approaches) is creating consistent command intent across the relevant set of organizations and functions involved. Note the use of the term *consistent* as opposed to the stronger term *congruent* (which we and others have identified as ideal), which would imply that intent across the enterprise is identical. Given the variety of charters, agendas, and cultures (organizational and national) involved in any reasonably sized mission, the best that can be achieved in practice is consistency of intent. This point has also been made by those who seek to replace the classic principle of war "unity of command" with the principle "unity of effort." Even unity of effort is, from what we have seen in Bosnia, Rwanda, Kosovo, Afghanistan, Iraq, and humanitarian assistance efforts around the world, an ideal that is seldom achieved in practice. We have concluded that what is achievable in coalitions, humanitarian, reconstruction, and peace operations is *unity of purpose.*

Indeed, it is often a sign of progress if agreement on purpose allows the different organizations dealing with a situation to de-conflict their efforts (e.g., not all deliver water or food to the same locations) and very valuable if they are able to develop synergistic[57] relationships. In situations where competent organizations with very different organizational perspectives

---

[57] For more on *synergy*, see:
Alberts et al., *Understanding Information Age Warfare.* pp. 211-212.

Allocation of decision rights

come together to deal with complicated and difficult problems, the minimal Information Age approach might best be described as "constructive interdependence." In this concept, military organizations might be providing security, lift, and local communications services while international organizations take responsibility for collecting information about violations of human rights and properly identifying and registering refugees; non-governmental organizations worry about emergency supplies of food, water, medicine, and shelter; and the host government focuses on delivering local police services and ensuring that schools and medical facilities are made available to the population. Such efforts require unity of purpose, but they also make it possible for competent authorities to cooperate and collaborate intelligently and use their mutual dependencies as assets. In Industrial Age conceptions, those dependencies are bad—they reduce the control of the organizations involved. That is why they tend to seek de-confliction so that they will have all the assets needed to complete a task or mission and no other organization will interfere with their efforts. This leads to optimization at the task level, but actually prevents synergy. Hence, it limits overall performance.[58]

This discussion underscores the fact that decision rights are, *de facto*, distributed differently today in the vast number of operations involving the military than they have been in the past. Even in the "high stakes" efforts of the U.S. military, at this writing, in Afghanistan and Iraq, the U.S. counts heavily on coalition partners, works closely with international organizations, depends on private voluntary organizations for specialized services, and is present at the invitation of and

[58] Industrial Age "de-confliction" is discussed in:
Alberts and Hayes, *Power to the Edge.* pp. 37-51.

works with, rather than controls, a host government. Indeed, the long-term measure of success (a measure of policy effectiveness) in these situations is the ability of the military to (1) engage only in those functions necessary for success, (2) turn over more and more responsibility to other competent and legitimate actors—to further distribute decision rights, and (3) eliminate the need for continued U.S. involvement (or reduce the need to a minimal level).[59]

Even with this in mind, however, critics have noted that there are two positions in which some central authority needs to play a role. The first has already been mentioned: the need to develop consistent command intent across the communities involved. In a very real sense, this is a role that can and should be broadly distributed. The "consent of the governed" is enshrined in the theory of democracy and is a major factor in self-synchronization and in Edge organizations. In military situations, leadership plays an important role in deciding what needs attention, framing the issues, developing approaches that are both feasible and consistent with the values of the society (or societies) involved, and ensuring that the goals and fundamental approach are understood by all those involved. Even in specific engagements where self-synchronization is intended, leadership has a crucial role in ensuring that the elements of the force are well-prepared. The role of Admiral Nelson in the Battle of Trafalgar illustrates this well.[60]

---

[59] Military forces cannot escape their obligations under international law to provide basic services in occupied territories and may have moral obligations to provide humanitarian assistance (food, water, shelter, medical treatment) in a variety of situations, but they should always be seeking to ensure that these roles are turned over to other competent, legitimate organizations as rapidly as is practical and consistent with their legal status (e.g., the U.N. or other international organization mandate, legitimate request from foreign government for assistance).

[60] Alberts and Hayes, *Power to the Edge.* pp. 28-31.

The second crucial role for leadership (or military command) is establishing the capability to communicate (enabling patterns of interaction). At the level of creating and maintaining the enterprise, this can be accomplished by building the now proverbial "system of systems" or by establishing standards or protocols that enable the appropriate level and quality of communication, information exchange, and collaboration required for success. Note that this requires either a central decision (this is the set of interaction mechanisms we will use) or development of a consensus about the appropriate mechanisms.

Hence, the same goal (efficient and effective patterns of interaction and distributions of information) can be achieved with either centralized or distributed decisionmaking. However, that goal must be achieved if self-synchronization is to be enabled in other functional areas. For Information Age military establishments, this will occur through a combination of decisions involving both broad principles and standards. These principles and standards should be developed and implemented within the commercial marketplace. That practice will make it much easier for Information Age military organizations to share information and collaborate with non-military actors around the globe. The SPAWAR concept of *composable systems*[61] is a recognition of this requirement. These principles and standards will also need to provide information assurance in all of their key dimensions.

Note that when the time comes to employ the enterprise (or part of the enterprise), the military may be only part of the

---

[61] Galdorisi et al. "Composeable FORCEnet Command and Control: The Key to Energizing the Global Information Grid to Enable Superior Decision Making." Presented at the 2004 Command and Control Research Technology Symposium. San Diego, CA. June 15-17, 2004.

effort. In these cases, a new more temporary enterprise is being created (e.g., those responding to the 2005 Asian Tsunami or those bringing international relief to Rwanda) and some coherent decision must emerge about the mechanisms that will be used for information exchange and collaboration.

Information Age militaries must also be able to avoid chaos. Truly uniform decision rights, in which every individual is involved directly in every decision and has an equal voice in each one, is unlikely to work for most military missions. The obvious exceptions are very small, highly competent, and very capable units such as Special Forces. In their classic roles, they develop rich information about each mission, tailor the team and its equipment to maximize the probability of success, plan the effort in great detail (using peer review to test the concept of operations and the plan), rehearse the mission multiple times, and distribute responsibility for tactical decisions throughout the team.[62] Karl Weick and his colleagues also note that larger professional military (and other) organizations with complicated, high risk tasks (such as those manning the flight deck of an aircraft carrier or fighting forest fires) also delegate considerable responsibility to their junior members.[63] However, in all these cases, the people in the force understand:

- command intent;
- relevant courses of action;
- how they map into different possible circumstances;
- that there is a high level of trust present in the organization; and

---

[62] Booz Allen Hamilton. "Network Centric Operations Case Study: Naval Special Warfare Group One (NSWG-1)." 2004.
[63] Weick, K.E. & K.M. Sutcliffe. *Managing the Unexpected: Assuring High Performance in an Age of Complexity*. San Francisco, CA: Jossey-Bass. 2001. p. 109.

- that those involved are highly competent.

These (and some other factors related to information distribution and patterns of interaction) appear to be the minimum conditions necessary for both the broad distribution of decision rights and effective performance.[64]

Finally, the distribution of decision rights is crucial to defining the communities of interest involved in complicated efforts like warfighting, national reconstruction, peace operations, or humanitarian operations. A community of interest (COI) is composed of all of the actors who care about and can influence the decisions made on a particular subject.[65] For example, the logisticians supporting a force actually form a community of interest, as do those responsible for targeting or for air defense. Often the distribution of decision rights is best assessed by looking at how they are distributed within the COIs that perform the effort. There is good theoretical and empirical evidence that broad participation in decisionmaking increases the quality of the decisions made.[66] At the same

---

[64] Both are necessary once the goal is not simply to be an Edge organization, but to be a successful organization.

[65] DoD defines a COI as "any collaborative group of users who must exchange information in pursuit of their shared goals, interests, missions, or business processes." Department of Defense, CIO. "Communities of Interest in Net-Centric DoD, Version 1." p. 3.

[66] Additional insight can be gained from:

Janis, Irving. *Groupthink: Psychological Studies of Policy Decisions and Fiascoes*. Boston, MA: Houghton Mifflin. 1982.

Druzhinin, V. V. and D. S. Kontorov. *Decision Making and Automation: Concept, Algorithm, Decision (A Soviet View)*. Moscow, Russia: CCCP Military Publishing. 1972.

Brown, Rupert. *Group Processes: Dynamics Within and Between Groups*. Malden, MA: Blackwell Publishing. 2000.

Hayes, Richard E. "Systematic Assessment of C2 Effectiveness and Its Determinants." Presented at the 1994 Symposium on Command and Control Research and Decision Aids. Monterey, CA. June 21-23, 1994.

time, all organizations large enough to deal with the complicated and dynamic problems involving today's military forces are actually made up of overlapping communities of interest.

Here again, the Industrial Age model of isolated communities (e.g., logistics, intelligence, plans) is both unrealistic and also enormously inefficient and, particularly in unfamiliar situations, ineffective. It is precisely the interconnections among COIs enabled by Information Age technologies and processes that make synergy and effects-based operations possible. Far more emphasis needs to be placed on these interconnections so that semantic interoperability can be achieved, which in turn facilitates shared awareness and ultimately shared understanding and synergistic actions.

However, there may be a price for this broad distribution. All other things being equal, broad participation can mean slower decisionmaking under some circumstances, but this does not have to be true. "Hard" groups, those who have worked together before on similar problems, create organizational artifacts (specialized language, work processes, a repertoire of previous decisions they can reference in their dialogue), and operate much faster than newly formed groups. This, of course, is why professional military organizations stress training and exercises. If done well, they serve to build trust, develop a common language, and create common or compatible processes. Hence, COIs that have worked together before are better equipped for Information Age **C2**.

Moreover, for simple problems or situations (those that are familiar, have a finite number of well-understood alternative actions, and for which the correct action can be selected by rule or algorithm), military organizations develop doctrine[67] or simple tactics, techniques, and procedures (TTPs) that guide effective execution of particular tasks. Doctrine and TTPs enable rapid, high quality decisionmaking in those situations for which they were developed and also bring consistency to the decisionmaking, making the behavior of one part of the force predictable and understandable by the other parts. Such doctrine and TTPs are, of course, an example of "pre-real-time thinking" in which an organization identifies situations that it expects to confront (e.g., crossing a river or a minefield) and develops the equipment and processes needed for success, and then builds them into its training and planning. Here again, the performance of the English fleet under Nelson at Trafalgar is instructive. He and his captains were intimately familiar with the capabilities of their own ships and crews as well as those of the adversary. They were familiar with the tactics of the enemy and the ones they wanted to use against them. They also met several times to discuss the battle plan.[68]

Patterns of interaction: Social domain

The basic work on patterns of interaction is discussed in some detail in *Understanding Information Age Warfare* (2001). Three key elements are specified for Information Age networks:

- Reach (the number and variety of participants),
- Richness (the quality of the contents), and

---

[67] The Department of Defense defines doctrine as: "Fundamental principles by which military forces or elements thereof guide their actions in support of national objectives. It is authoritative, but requires judgment in application," rather than dogma to be followed to the letter even when circumstances make its application impractical or unwise.

[68] Alberts and Hayes, *Power to the Edge.* pp. 28-31.

- Quality of interactions enabled.

The first and third of these are crucial for the patterns of interaction in any future C2 Approach. Together they provide important insights into the C2 Approach of an enterprise: who is "on the net," what is the quality of their information, and how well can they collaborate?

Those who have looked at patterns of interaction often inappropriately limit their view to focus only on connectedness:[69] who is linked to whom within a military force, in a coalition, in an interagency process, or in a broad international effort. While this is certainly part of the idea, understanding patterns of interaction requires focusing on more than just connectivity needs. First, the level of interoperability achieved cannot be ignored. Interoperability means not only technical interoperability, but also semantic interoperability (the capacity to fully comprehend one another) and "cooperability" or willingness to interact and desire to communicate clearly.

Second, the range of media across which these interactions occur is also crucial. Voice connectivity over an unreliable system such as HF radio does not enable nearly as rich a set of interactions as video conferencing, the ability to exchange email, or shared whiteboards that can be used to discuss a rapidly evolving situation and the relevant courses of action available.

Third, the most desirable patterns of interaction are collaborations (working together toward a common purpose), as opposed to interactions that only involve the exchange of data or information. Collaboration provides the opportunity for the parties

---

[69] A minimal concept of reach.

to exchange views about the clarity of the data and information, as well as what it means or implies, not just to receive them.

Fourth, digital connectivity is qualitatively better than pure voice. Voice alone means that the speaker must be clear, the transmission understandable, the listener must be paying attention and able to hear everything, and the message captured correctly. Even small errors in any of these processes (transpositions of coordinates, missing key words, etc.) in communication can lead to meaningful differences in shared information, shared awareness, shared understanding, decisionmaking, or action synchronization.

The mechanisms by which information is exchanged may also vary across **Command and Control** Approaches. Traditional **C2** Approaches relied heavily on *information push*:[70] the originator of the information was responsible for deciding what to share, how to organize or format it, to whom to send it, and how often to update the information. This required elaborate planning, detailed standard operating procedures, and specific protocols. In many cases, it also required that the recipient acknowledged receipt. It always meant that the recipient had to ask questions if what was received was unclear or garbled, and to take the initiative if expected information (e.g., a status report or an intelligence update) was not received or if some non-standard information was needed or desired.

As satellite and other systems that provided increased bandwidth became available, some items of general interest (for example, weather in an area of operations) moved to *broadcast*

---

[70] Information push occurs when one person sends to another information that they think the other person needs.

media. Here again, the originator was responsible for deciding content and when to update the information. When the mechanism for updating information is broadcast, the user must be synchronized in time to receive the information (unless the broadcast is recorded for playback at a convenient time). Subscription services are more selective as information is not merely broadcast into a "black hole" but is sent to a collection of subscribers who have a need for or express interest in some specific information. Thus when an individual obtains a certain piece of information, he knows who needs or wants it, rather than attempting to assess who needs what.

With digital technologies and networks available, the originator of data can take a different approach to making information available. He is free to *post* the information in a space that is available to those consumers who have a legitimate need for it. While some information needs to be secure and the rights to alter information need to be controlled (as well as providing pedigrees and the other elements of information assurance), digital networks provide the opportunity for a large number of users—and users that the originator did not know needed the information—to access it once it is posted. This approach also enables information to be made available in a more timely manner.

Posting information makes it necessary for consumers to know what has been made available and how to acquire it. This requires both education and tools. Moreover, when information is posted, the issues of information quality and authenticity become extremely important. Indeed, this type of information sharing allows the users to *pull* information, which is a genuinely new method of interaction. Increasingly, we are seeing references to the concept of "smart pull," which assumes an empowered user with both the capacity for intelligent search and discovery, as well as the capability to bundle the volumes of relevant information. Information consumers must also be able to accomplish this "pull" within the time available.

The richest possible pattern of interaction would be facilitated if there was a broadband system that digitally linked every entity to every other entity, had full interoperability (technical, semantic, and cooperability), and provided a cyber environment that supported continuous collaboration. At the other end of the spectrum would be a system where each actor can only talk with one, or perhaps two others (a superior and a subordinate), their interactions are very constrained (little bandwidth), and interoperability is low among some or all of the constituent elements. The types of transactions permitted and encouraged (push, post, pull, or subscribe) must also be understood if patterns of interaction are to be understood.

Industrial Age patterns of interaction

Within Industrial Age organizations, the patterns of interaction are designed to ensure control from the center. Hence, the flow of information follows the "chain of command" or the management structure of the enterprise. We can see this pattern in the rule for exchanging information within the U.S. military. All official correspondence is addressed to the commanding officer of the unit, with the specific person who needs the information identified in an "attention" line. This practice reinforces the long-established tradition that information must flow along command lines and that all information within a command is the business of and belongs to the commanding officer. Hence, Industrial Age patterns of interaction mimic the hierarchical structure of the organization.

In many cases, limited bandwidth for communications and a high volume of traffic has led to not one but several sets of hierarchies in Industrial Age militaries. For example, different radio nets are used for command, artillery fire control, control of air assets, and logistics. Another factor that affects the flow of information is that each of the military services has unique communications systems. In addition, the intelligence community has specialized communications systems that are more secure and limit access to those with appropriate clearances and a need to know. The problem is that these specialized hierarchies involve very few horizontal linkages, which makes them weak sources of collaboration. These information flow patterns lie behind the ubiquitous criticism of "stove pipes" in Industrial Age militaries. They both reduce the flow of information and slow it by forcing it to pass through a number of intermediate points, the layers of middle management.

Enabling the linkage of "everyone to everyone" does not imply that a direct connection exists between all pairs of entities, or that everyone talks with everyone else. It does mean that, if needed and appropriate, anyone can exchange information and interact with anyone as necessary. The Industrial Age instantiation of such a system was the telephone system, which enabled people to talk around the world, but did not provide (and lacked the capacity to provide) open lines between every pair of telephones around the globe.

Information Age patterns of interaction

Patterns of interaction are actually networks. In the case of NCW and Edge organizations, these are social networks that will be enabled by whatever mechanisms are available: courier, telephone, videoconference, local area networks, wide area networks, the World Wide Web, etc. Social networks also depend on cooperability: the willingness to work together and collaborate when appropriate. The study of interaction networks is an important area of science, research, and development.[71] As the importance of networks in complex structures and complex adaptive systems has become recognized and tools such as graph theory have emerged to help mathematicians and scientists understand their properties, some significant insights for Information Age patterns of interaction have emerged.[72] In many cases, these insights arise from observation and analysis of networks in the real world, ranging from physical interactions in natural systems to social networks involving humans.[73] In other cases, they have arisen from academic efforts to understand networks and their important characteristics.

For purposes of examining approaches to command and to control in Information Age militaries, we need to distinguish four different types of networks, none of which is a hierarchy:

---

[71] A thoughtful introduction into social networks can be found in:
Wasserman, Stanley and Katherine Faust. *Social Network Analysis: Methods and Application*. Cambridge, UK: Cambridge University Press. 1999.
Scott, John. *Social Network Analysis: A Handbook*. London, UK: Sage Publications, Ltd. 2000.
Degenne, Alain and Michael Forsé. *Introducing Social Networks*. Thousand Oaks, CA: Sage Publications, Ltd. 1999.
Gladwell, Malcolm. *The Tipping Point: How Little Things Can Make a Big Difference*. United States: Little, Brown and Company. 2000.
[72] For more on Information Age patterns of interaction, see:
Barabasi, Albert-Laszlo. *Linked: How Everything Is Connected to Everything Else and What It Means*. New York, NY: Penguin Group. 2003.
Buchanan, Mark. *Nexus: Small Worlds and the Groundbreaking Science of Networks*. New York, NY: W.W. Norton & Company, Inc. 2002.
[73] Johnson, Steven. *Emergence: The Connected Lives of Ants, Brains, Cities, and Software*. New York, NY: Touchstone. 2001.

- Fully connected networks,
- Random networks,
- Scale-free networks, and
- Small world networks.

Fully connected networks, in which every entity directly interacts with (or is connected to) every other entity, have been explored in detail in the small group literature.[74] In small group research (largely laboratory experiments and field studies), fully connected networks have proven effective at solving complex problems, but are often slower than more streamlined structures such as hierarchies or spokes of a wheel around a central person, particularly when the groups have no prior experience working together.[75]

For military systems, however, the important fact is that fully connected social networks do not scale well. For a fully connected network that consists of N nodes, every node that joins adds N-1 linkages. This is costly to support, quickly overloads any available bandwidth, and means that the number of interactions possible for any node rapidly becomes overwhelming. Hence, even if such a system is constructed, the nodes within it must make an enormous number of decisions about when they will interact, with whom they will interact, and how much attention they will pay to any interaction or offer for an interaction. Hence, the endpoint on the spectrum of patterns of interaction—everyone with everyone else, all the time, and using the full range of media—is simply impractical as an approach for a large-scale military organization or an interna-

[74] An introduction into the field of small group literature can be found in: Fisher, B. Aubrey. *Small Group Decision Making: Communication and the Group Process*. New York, NY: McGraw-Hill Book Company. 1980.
[75] Druzhinin and Kontorov, *Decision Making and Automation*. 1972.

tional effort. That does not mean that fully connected social networks cannot be useful in small operations. In fact, the "clusters" that are valuable in Edge organizations within other types of networks may well be fully connected themselves. Indeed some meaningful research into teams and collaboration[76] indicates that rich connectivity can be very beneficial and may be necessary for success.

A *random* network develops when each node has an equal probability of interacting with any other node. The distribution of interaction in random networks can be expressed as a normal or "bell" curve, though it is in fact a Poisson distribution when the interactions are relatively rare.[77] Because of the underlying random properties, these networks are sometimes referred to as "egalitarian" networks. However, these networks are not very efficient—it may well take a very large number of steps or linkages to move from one node to another. This property changes as the size of the network (number of nodes) and the relative density of the interactions between nodes change, but it will consistently form a bell curve if the network is truly random. Random networks will also form relatively few clusters. In other words, while they have low average path length, they also have a low clustering coefficient. Hence, while a great deal of mathematical and theoretical work has been done on random networks, and some on random social networks, they have relatively little practical utility in **C2** system design or implementation.

---

[76] Noble, David F. "Understanding and Applying the Cognitive Foundations of Effective Teamwork." Prepared for the Office of Naval Research by Evidence Based Research, Inc. April 23, 2004.
[77] Atkinson, Simon Reay and James Moffat. *The Agile Organization: From Information Networks to Complex Effects and Agility*. Washington, DC: CCRP Publications Series. 2005. p. 14.
Buchanan, *Nexus*. 2002.

Except when they are very dense and approach a fully connected network, random social networks lack resilience, the capability to persevere despite obstacles or setbacks. Because only a modest percentage of their interactions occur across any one area of the space in which they are located, removing a modest percentage of the nodes or linkages in a random network will result in it splintering into a number of unconnected structures. Hence, they are quite vulnerable to attacks and may degrade quickly if linkages are sparse.

Scale-free networks are characterized by an extreme distribution of interactions among their nodes. A few nodes have a very large number of interactions. Most nodes have very few interactions. The term *scale-free* comes from the fact that this distribution is so extreme that it approaches an exponential distribution. (Actually, exponential distributions go to zero while scale-free networks go to a very low number, but continue their tails well beyond where an exponential distribution would reach zero.) More correctly, scale-free networks have a power law distribution.[78] Rather than forming the classic bell shape, this distribution is sharply skewed toward the origin, with a long flat tail. Scale-free networks are found throughout nature,[79] wherever complex adaptive systems develop. Examples include the distribution of branches in a river system and the distribution of nodes in the Internet. These networks are efficient (only a few steps are needed to move from any one node to another). They are also resilient, much more so than random networks. However, they can be vulnerable if an adversary knows how to find the key nodes and the linkages between and among them.

---

[78] Albert, Reka and Albert-Laszlo Barabasi. "Statistical Mechanics of Complex Networks." *Reviews of Modern Physics*. Vol 74, No 1. 2002.
[79] For additional information on naturally occurring networks, see: Buchanan, *Nexus*. 2002. & Johnson, *Emergence*. 2001.

The fundamental difference between random networks and scale-free networks is the existence of very long linkages that reduce the number of steps required for an interaction to move from one part of the space to another and a set of naturally occurring clusters throughout the system. These clusters are often described as "hubs" because they will tend to form around one node that is involved in one or more long-haul linkages. Even a few such linkages, provided that they link nodes that serve as hubs for clusters within each region of the network, create a remarkably strong and capable social network. Most stable, naturally evolved complex adaptive systems have been shown to operate as scale-free networks. Because of the importance of the key nodes involved in these long linkages, they are sometimes referred to as "aristocratic" networks in contrast to their random, egalitarian cousins.

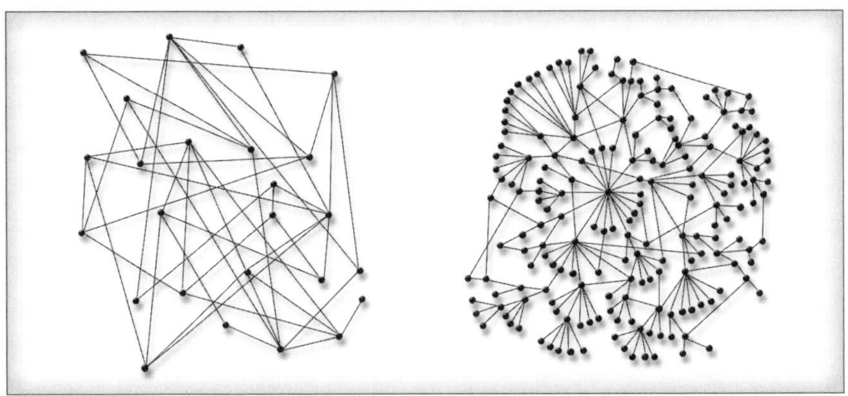

FIGURE 14. COMPARISON OF A RANDOM NETWORK (LEFT) AND A SCALE-FREE NETWORK (RIGHT)

The richest, most efficient class of network currently under study is the *small world* network.[80] These systems got their name from the discovery that moving information from one

80 For more on small world networks, see: Watts, *Small Worlds.* 1999.

part of the network to another requires only a small number of steps.[81] The distinguishing feature of small world networks, however, is a very large clustering coefficient. As a result, a link to any one node is readily tied to a number of other nodes. This is the actual structure of most effective groups of experts. They are, in one sense, random social networks growing toward scale-free social networks but not yet characterized by the number of long-haul linkages that would transform them to meet the definition of scale-free. In other words, these small world networks link together clusters that can be thought of as communities of interest or communities of practice. Each such community forms its own "small world" which may be linked to other small worlds by some individual nodes that act as connectors. These small world networks can evolve from or evolve into scale-free networks. However, if they are temporary (for example the "hastily formed networks" that arise to deal with the aftermath of a natural disaster or to carry out a specific, temporary military mission), they will function relatively efficiently and robustly while needed, then gradually disappear as the interactions become less frequent and the linkages atrophy or are abandoned.[82]

---

[81] Stanley Milgram executed a small world network experiment in 1967. The experiment revealed that two random U.S. citizens were on average connected via six acquaintances. Hence the common term, "six degrees of separation." Milgram, Stanley. "The small world problem." *Psychology Today*. Vol 2. 1967. pp. 60-67.

[82] Brian Steckler of the Naval Postgraduate School, among others, established a hastily formed network in Thailand after the December 26, 2004 tsunami that struck the Indian Ocean region. This wireless network proved to be critical in supporting the relief and recovery effort in the area of operation. Also see:
Honegger, Barbara. "NPS IT Fly-in Team Reconnects Tsunami Survivors to the World." Naval Postgraduate School. Feb 9, 2005.
Owen, Don and Eugene K. Hopkins. "Southeast Asia Tsunami Research Project." Report commissioned by the CCRP and prepared by Evidence Based Research, Inc. 2005.
Comfort, Louise K. "Risk, Security, and Disaster Management." *Annual Review of Political Science*. Vol 8. 2005.

The networks required for self-synchronization or Edge organizations are small world networks in which those with the relevant knowledge and capabilities form richly linked and frequently interacting clusters that permit them to exchange information, develop shared situation awareness, and collaborate in order to synchronize their plans (explicit or implicit) and undertake synergistic actions.

Richest network structure

The richest and most resilient network structure (the pole for the distribution) appears to be a hybrid that looks at the global level like a scale-free network, but at the intermediate level is composed of small world networks, and at the local level fully connected social networks. This combination appears to provide the blend of efficiency, effectiveness, and resilience needed for large-scale enterprises operating in multi-dimensional, dynamic environments. These patterns of interaction are capable of becoming complex adaptive systems. Of course, like other ideal types, this hybrid form does not exist today.

Distribution of Information

The term *information* here covers a range of phenomena including data, information, understanding, knowledge, and wisdom. We have defined these terms elsewhere, as have other researchers.[83] Recent work of the SAS-050 NATO Working

---

[83] For more on data, information, understanding, knowledge, and wisdom, see: Alberts et al., *Understanding Information Age Warfare*. pp. 16-17, 19, 95-117.
Ackoff, Russel L. "From Data to Wisdom." *Journal of Applies Systems Analysis*. Vol 16. 1989. pp. 3-9.
Zeleny, M. "Management Support Systems: Towards Integrated Knowledge Management." *Human Systems Management*. Vol 7. 1987. pp. 59-70.

Group has stressed that data, when placed in context such that it reduces uncertainty, becomes information, while information becomes awareness when it passes from information systems into the cognitive domain (a human brain). Humans, as individuals, actually hold awareness of situational information and combine it with their prior knowledge and mental models (which include perceptual filters that may prevent full awareness of some information) to generate situation understanding, which includes some perceptions of the cause and effect relationships at work and their temporal dynamics. As was discussed earlier, these elements of the sensemaking process also drive decisionmaking. We speak of *shared information* as that directly available to more than one actor in a social network, and *shared awareness* and *understanding* as those elements that are common across more than one entity or node.

This distribution of information dimension refers to a key result of the **C2** processes within a military organization, coalition, or international effort (e.g., reconstruction, peace enforcement, humanitarian assistance) involving military forces and civilian organizations, which might be other government agencies, international, or private entities. As noted earlier, the distribution of information is impacted by the distribution of decision rights (which includes who makes the choices about information distribution processes and the creation of the infrastructure by which information is shared and collaboration is carried out, as well as who is entitled to what information) and the patterns of interaction (who is able to acquire what information). However, the distribution of information also has a dynamic of its own. The concept of the distribution of information also includes the *richness* element in network-centric thinking. Richness focuses on the breadth,

Distribution of information

depth, and quality (correctness, completeness, currency, consistency, etc.) of the information that is available.[84]

The pattern of distribution of information within an enterprise arises partly from the allocation of decision rights, partly from the patterns of interaction, partly from the willingness to share information, partly from the willingness of individuals to acquire it, and partly from the tools and skills they have to acquire it. This includes and indeed is strongly impacted by their ability to collaborate, particularly when considering the more strongly cognitive aspects of information—understanding and knowledge—and the ability to share information, awareness, and understanding. Ultimately, the distribution of information governs the capacity for sensemaking at both the individual and collective levels.

As a simple dimension, the distribution of information can be thought of as ranging from fully centralized repositories (e.g., the old mainframe computer that held everything for a company or organization and the access of each user was predetermined and controlled by a central authority) to fully distributed (networked) approaches wherein everyone has access to everything, but storage is redundant.

Industrial Age distribution of information

The Industrial Age involved functional decomposition and role specialization. Hence, during that era information was distributed according to the specific needs of each user. This required preplanning and the identification of the "owner" of

[84] For more on the richness of information, see chapter six of the joint ASD NII/OFT work entitled: "NCO Conceptual Framework Version 2." pp. 56-66.

each information element. Owners were responsible for ensuring the quality of the information and for ensuring that it was distributed according to the organization's plans. Changes in the processes for collecting, developing, or distributing information required centralized decisionmaking, which would often be conducted by committees or groups of specialists. The engineers designing C2 systems spent a great deal of time developing information exchange requirements (IERs) that specified who needed access to what information under what circumstances. These IERs then drove which linkages were enabled with what capacity. More importantly, they also resulted in *de facto* decisions about who would not have access to information. In other words, they constrained interactions and limited the distribution of information under the Industrial Age assumption that they knew, with precision adequate to engineer these systems, the threats and circumstances under which military forces would operate.

Industrial Age militaries followed this practice of preplanning, centralized systems, and constrained distributions. For example, intelligence information was kept in intelligence channels, logistics information in logistics channels, etc. These functions came together at their common bosses: the military commanders. Those in the command roles could, and often did, mandate distribution across functions where integration appeared useful or wise, but unless such mandates were issued, information stayed in functional channels. Military echelons also acted as constraints on the distribution of information. Here again, command decisions (often in the form of standard operating procedures) were required to move information across echelons.

Distribution of information

The impact of these Industrial Age patterns of information distribution can be seen in the sequential nature of the decisionmaking processes. Figure 15 shows that this system of information distribution all but guarantees that all decisions are based on outdated information flowing upward and outdated guidance flowing downward. It also highlights the possibility for a lack of synchronization or synergy across functional areas such as intelligence, operations, and logistics.

Information Age distribution of information

As the capacity for sharing information has grown, the distribution of information has tended to break away from rigid guidelines based on function and echelon. Similarly, where the capacity for collaboration has emerged, the quality of information has improved (people with different perspectives and different knowledge or expertise have an opportunity to comment on information and compare it with other information, understanding, and knowledge) and the patterns of distribution have been enriched. These processes are well underway in some aspects of today's forces, but are still a very long way from the "robustly networked force" envisioned in NCW.

In a genuinely Information Age enterprise (or an Edge organization), all information is available to all the entities, with constraints minimized and focused on necessary aspects of information assurance (privacy, integrity, authenticity, availability, and non-repudiation). However, information assurance is not adequate in itself. Users must also have the tools to find relevant information (discovery and search capacity), a rich understanding of what information is available, and the capability to process it so they can "digest" it or use it to add to their sensemaking.

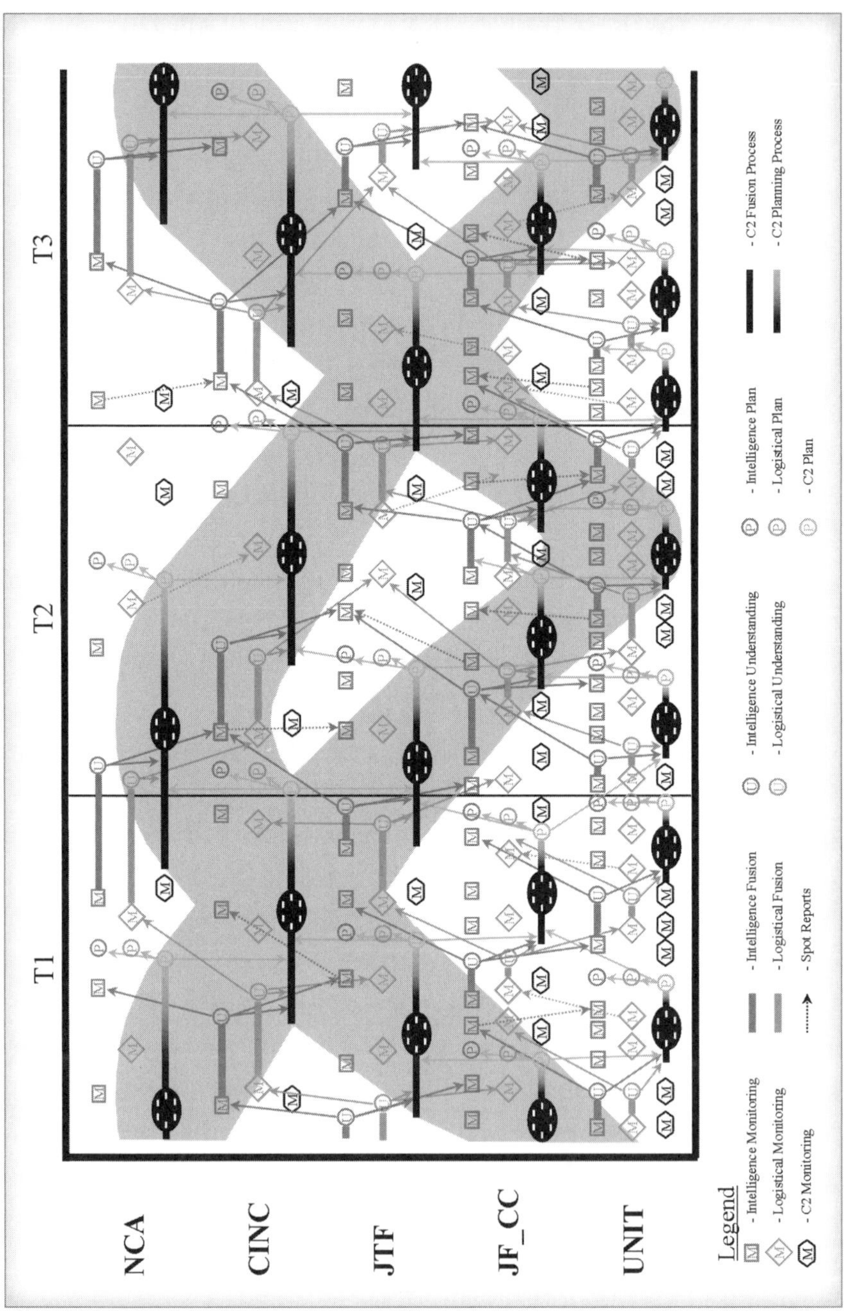

FIGURE 15. INDUSTRIAL AGE INFORMATION FLOW

Distribution of information

Information assurance is also, by itself, an inadequate characteristic for Information Age information distribution because it does not take into account the need for agility. For example, it needs to consider the distribution of storage as a defense against loss of connectivity from malicious attacks, natural disasters, or systems disasters, as well as the need to rapidly redistribute information in dynamic or unforeseen developments. Both robustness of the system and redundancy are key elements for maintaining effective information distribution over time and across function.

Conclusion

The core of understanding **Command and Control** lies in its definition and the three key dimensions by which alternative **C2** Approaches can be characterized and differentiated: the allocation of decision rights, the patterns of interaction that occur, and the distribution of information. These apparently simple dimensions are interrelated and drive both the command function and the control function. Together, they define the **C2** space.

The chapters that follow put these key dimensions under the lens of a "value view" (Chapter 7) and "process view" (Chapter 8), both to highlight their implications and also to provide examples of how they can be expected to play out in forces, coalitions, interagencies, and public-private partnerships involving the military.

Chapter 7

Conceptual Model: Value View

The value view of a conceptual model focuses on the relationships between and among selected[85] measures of merit (MoM) associated with the model's functional or process concepts. The boxes represent measures of merit related to these concepts while the arrows represent the existence of a relationship between the measures. It is important to understand that the relationships depicted in the value view are, in a very real sense, integrals over situations and circumstances. These relationships (general relationships, e.g., when A increases then B increases, and specific relationships, e.g., a parametric function) represent general rules, propensities, or tendencies. In individual cases (a given mission or set of circumstances), a particular relationship may be stronger, weaker, or may, indeed, not hold at all.

---

[85] Because virtually all of the variables identified as relevant by their inclusion in a conceptual framework or model can, for a particular problem formation, be dependent variables and because, in many cases, dependent variables are considered to be MoMs, the value view presented in this chapter involves only a subset of MoMs or potential MoMs. Those that were chosen to be included (1) represent the major attributes of mission capability packages and (2) are key value-related concepts that constitute a mission value chain.

If the possibility that some relationships, even important relationships, may not always hold, then one might ask, "what good is a general value view?" The answer lies in the significant variations in missions and circumstances and the difficulty of predicting which missions we will undertake and the circumstances that will be associated with a given mission. Given the level of uncertainty about the future and the fact that virtually all of the organizations and the materiel and systems in which we invest have multiple uses, a value view provides a crucial anchor for defense planners making decisions about how to shape the force, what characteristics our organizations should have, what **C2** Approach we should take, and what capabilities the force should have.

Without a value view, the only way to inform these policy and investment decisions is to make some specific assumptions about missions and circumstances in the form of "planning scenarios," to use these scenarios to instantiate our process model, and to employ tools to create data that, in turn, can be analyzed. The significant shortcomings of this approach, even for analysis of more traditional military missions, are well-recognized.[86] As a result, DoD policy now calls for a capabilities-based planning approach rather than a threat-based one.[87]

Agility of the force has been an essential characteristic for 21st century organizations in general and militaries in particular.

---

[86] Khalilzad, Zalmay M. and David A. Ochmanek. *Strategy and Defense Planning for the 21st Century*. Santa Monica, CA: Rand. 1997. pp. 71-85.

[87] Unlike the threat-based approach of the past, capability-based planning focuses on how an adversary may challenge the U.S. instead of who the adversary is and where they may be engaged geographically. Although this change was being addressed prior to September 11th, the attacks certainly galvanized the need to change the planning process. Quarterly Defense Review Report. U.S. Department of Defense. September 30, 2001.

The value view, as an integral over a wide variety of missions and circumstances, offers the opportunity to understand the specific characteristics that are associated with agility and hence to shape a force with agility in mind.

When trying to examine new **Command and Control** Approaches and the nature of coevolved mission capability packages that employ such approaches, empirical data is in short supply. We can try to understand the relationships among the elements of such mission capability packages by instantiating them for a wide variety of missions and circumstances. This would take a very long time and make it necessary either to forgo force shaping and investment decisions that involved a move to new approaches or to commit to new approaches without evidence. This is clearly not an enviable position for decisionmakers. The alternative is to construct a value view based on all available evidence and, until more evidence becomes available, to fill in the gaps with reason and sensitivity analyses.

Careful observation of ongoing operations, instrumentation of exercises, the conduct of campaigns of experimentation, and the development and use of mission-specific process views of the conceptual model will, over time, provide a wealth of empirical evidence and analytic results. As additional evidence and analytical results become available, the value view provides a useful way to organize the evidence and analytical results, and provides a scientific basis for either making changes to the conceptual model or annotating it. Some relationships may be confirmed, and others annotated to note the conditions under which they appear to break down.

Value view: Overview

As we left it in Chapter 5 (Figure 10), the value view includes some measures of value that reflect, in one way or another, the contributions made by information and communications capabilities. There is a well-recognized need for coevolution, that is, for simultaneous changes to all of the elements of a mission capability package. For example, as access to information improves, individuals and organizations will be in increasingly better positions to make decisions and do things they do not do today because they must rely on others with better access to information to make certain decisions for them. This, of course, takes time and as a result units are not as responsive as they could be. If access to information is improved without coevolving the decision topology and processes, then the organization will not take advantage of the improved access to information.

Figure 16 adds to Figure 10 three measures of quality associated with the elements of MCPs: quality of personnel, quality of training (which for our purposes includes education), and quality of materiel. Without adequate attention to coevolution, we will not, as the example illustrates, be able to fully leverage our growing information-related capabilities nor take full advantage of the opportunities afforded by new **C2** Approaches. *Power to the Edge*[88] discusses both traditional **C2** Approaches as well as Information Age **C2** Approaches. This book also argues that it is unlikely that one **C2** Approach will be dominant over the full range of missions and circumstances and that it is likely that different **C2** Approaches would be

[88] A healthy discussion of Industrial Age **C2** and its shortcomings can be found in: Alberts and Hayes, *Power to the Edge*. pp. 37-53, 53-71, 201-213.

appropriate for different organizations, functions, or at different points in time during the same operation. This would require a dynamic ability to coevolve.

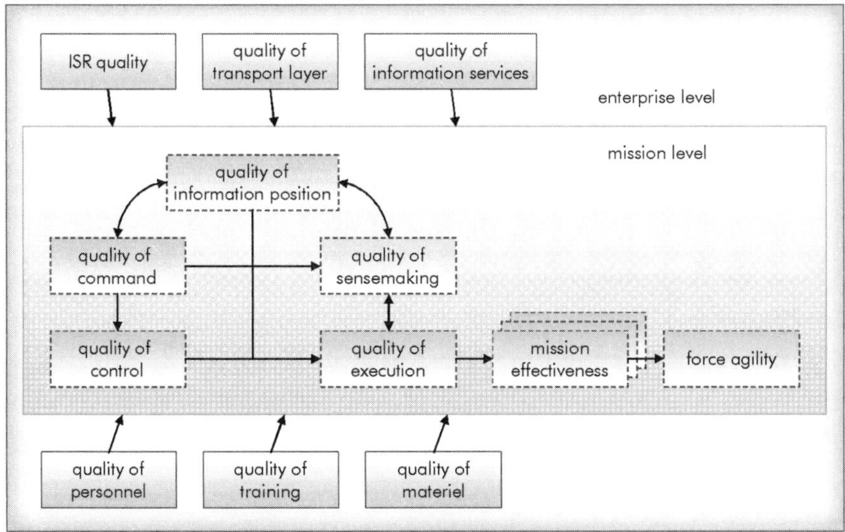

FIGURE 16. MCP VALUE VIEW

Understanding Information Age Warfare, in its discussion of the role of planning in synchronization, compares Nelson at Trafalgar with the plans used by the Egyptians in 1973, the former being an example of a flexible plan, the latter an inflexible one.[89] Changing the nature of a plan is not enough. For an approach to be successful, it must be appropriate to the nature of the mission capability package at hand. Nelson's approach, without the shared knowledge and training that went along with it, probably would not have been successful.

Figure 16 also changes *information quality* to *quality of information position*. This is because the value of the available information,

---

[89] See: Alberts et al., *Understanding Information Age Warfare*. pp. 219-221.

from a mission or enterprise perspective, depends not only upon the characteristics of the information available but also upon how it is distributed and the nature of the information-related interactions that can take place. This is discussed in detail later in this chapter.

THE NATURE OF THE FORCE

Mission capability package is one of the two terms that are commonly used to represent the set of materiel and non-materiel elements that constitute a deployable or fielded capability. In addition to MCP[90] (a mission capability package includes an operational concept and the force structure, C2 Approach, doctrine, education and training, and systems needed to make this concept a reality), the term DOTMLP[91] (doctrine, organization, training, materiel, leadership, and personnel) is in widespread use. Both terms were, at the time of their coining, reactions to a pervasive over-emphasis on materiel solutions. Both terms are meant to be inclusive, but their expressions emphasize different aspects of the capabilities that are required to conduct an operation. We prefer the term *MCP* to the term

---

[90] A discussion of mission capability packages can be found in:
Alberts, "Mission Capability Packages."
Alberts, *The Unintended Consequences of Information Age Technologies.* pp. 50-52.
Alberts, *Information Age Transformation.* pp. 74-78.

[91] Originally, the U.S. Army used DTLOMS (Doctrine, Training, Leader Development, Organization, Materiel, and Soldier Support) as the guiding model in order to develop an effective force. However, the term has evolved into DOTMLPF. Both models require a proper balance of the variables to achieve the ideal force.
Kelly, Terrence K. "Transformation and Homeland Security: Dual Challenges for the U.S. Army." *Parameters.* Carlisle, PA: U. S. Army War College. 2003.
DOTMLPF is at the heart of the transformative Joint Visions:
U.S. Department of Defense, *Joint Vision 2010.* 1996.
U.S. Department of Defense, *Joint Vision 2020.* 2000.
Filiberti, Edward J. *How the Army Runs: A Senior Leader Reference Handbook, 2003-2004.* Carlisle, PA; U.S. Army War College. 2003.

DOTMLP because it is inherently flexible and does not by virtue of its definition exclude important capabilities. Given that **C2** is arguably central to the conduct of military operations, any mission capability package should explicitly address the approach to both the functions associated with command and those associated with control.

To explore new **Command and Control** Approaches, the conceptual model needs to explicitly represent those elements of a mission capability package that most directly affect **C2** and the value view needs to explicitly represent the value of these elements. For this reason, three value metrics representing different capabilities provided by our C4ISR systems are included in Figure 16 where the original formulation of MCPs refers only to "systems." To facilitate discussion among those who may prefer using different terms for the force capabilities available or required for a given mission, a set of missions, or the force, the following table is provided.

MCP Elements	**DOTMLP**	**Value View**
C2 Approach	Doctrine, Organization, Leadership	Quality of Command, Quality of Control, Quality of Sensemaking
Training	Training	Quality of Training
Education	Personnel	Quality of Personnel
Systems	Materiel	Quality of ISR, Quality of Transport, Quality of Information Services
		Information Quality
		Quality of Execution
Mission Effectiveness	Mission Effectiveness	Mission Effectiveness
		Force Agility

TABLE 1. MCPS, DOTMLP, AND THE VALUE VIEW

Three major differences exist among these different ways of organizing the capabilities required to conduct an operation. First, MCPs and DOTMLP are defined in a way that reflects the characteristics of the force. These schemas or approaches are essentially descriptive. The value view takes these characteristics and translates them into expressions of value. Second, only the value view identifies information as a separate resource. In the DOTMLP schema, information would be included as part of materiel; in the MCP approach, information would be included in systems. Given the central role that information (its quality and distribution) plays in determining the quality of awareness and shared awareness, we believe it appropriate to focus increased attention on information. Third, both the MCP and the DOTMLP approaches are focused on describing the initial conditions prior to the employment of the force, while the value view is focused on variables that constitute a value chain leading to mission success or force agility. The value view considers the initial conditions insofar as they affect either the values of the variables that form the value chain or the relationships among them. Put another way, the value view provides an approach to understanding the implications of the characteristics of an MCP or a particular set of DOTMLP.

The remainder of this chapter is devoted to a more detailed discussion of each of the major concepts that are included in the value view. This discussion begins by looking at the quality of information position.[92]

[92] Information position: the state of an individual's information at a given point in time. See: Alberts et al., *Understanding Information Age Warfare.* p. 106.

Quality of Information Position[93]

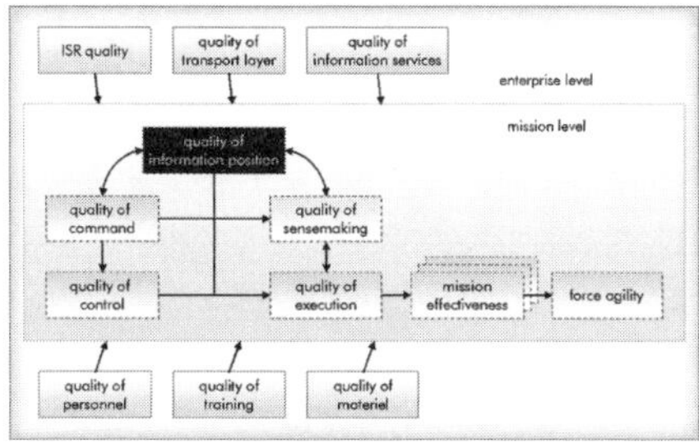

The word *information* is used by many to include data, knowledge, understanding, and even wisdom. These distinctions are not always important, however, to understand the conceptual model, particularly the value chain represented in the value view, these distinctions are quite important. Data are facts that when put into context become information. Information, to have value, must reduce uncertainty. Knowledge is more than archived information. Knowledge involves a transition from the information domain to the cognitive domain. It is a precursor to understanding. The quality of information position is in the information domain and hence is concerned about the quality of data and the ability to turn data into high quality information.

The quality of the information position needs to be considered at the individual, mission, and enterprise levels.[94] At any level, it is determined by more than the quality of the available sensors, reports, and analyses (data). It affects both the quality of sensemaking and execution.

---

[93] A substantial part of this discussion is based on material from *Understanding Information Age Warfare*, particularly Chapter 5. However, our thinking has evolved and thus this current discussion departs in some significant ways from previous work. Alberts et al., *Understanding Information Age Warfare.* pp. 95-117.

[94] The "mission level" encompasses teams, groups, or units, i.e. two or more individuals who share a common purpose. *Mission* is used here to include tasks, assignments, indeed any work performed collaboratively by two or more individuals.

This concept consists of multiple attributes (variables) that interact with one another. Figure 17 depicts its key components: information richness, reach, security, and the quality of interactions.

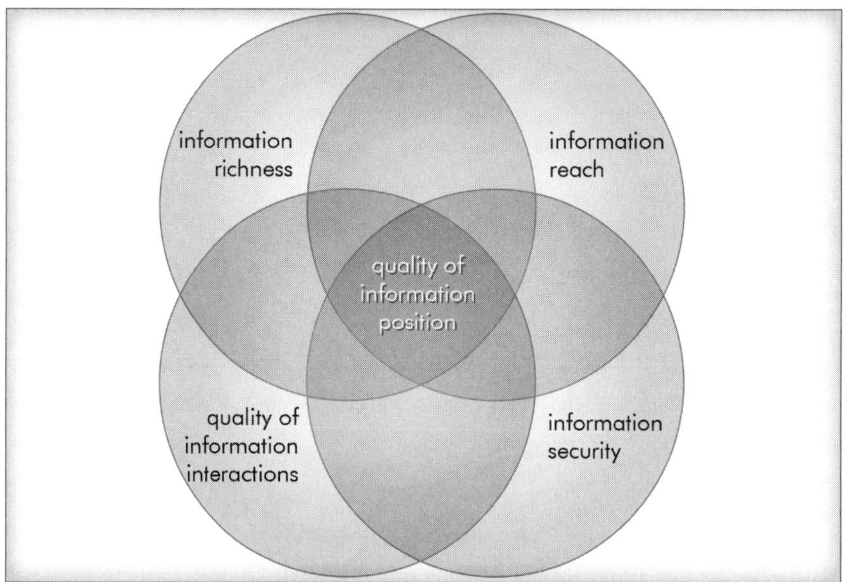

FIGURE 17. QUALITY OF INFORMATION POSITION

Information richness

Information richness consists of two sets of attributes: one that is independent of the situation (the individuals, the participants in a mission, and the enterprise) and one that is a function of the situation. There are four attributes of information that are situation independent—attributes that can be objectively measured.

- Correctness: the extent to which information is consistent with ground truth.
- Consistency: the extent to which a body of information is internally consistent.
- Currency: the age of the information.

- Precision: the degree of refinement, level of granularity, or extent of detail.

This set of objective attributes may be instantiated at the individual, mission, and enterprise levels. In the case of the mission and enterprise levels, the variables (correctness, consistency, currency, and precision) are reflective of the *union* of available information. That is, these measures are applied to the collective, the sum of the information across the individuals that are mission participants or, in the case of the enterprise, members of the enterprise.

The second set of attributes is a reflection of the utility of the information at *one* of the individual, mission, or enterprise levels. These variables measure the value of a particular attribute of information quality in a situational context.

- Relevance: the proportion of the information that is related to the task at hand.
- Completeness: the percentage of relevant information attained.
- Accuracy: the degree of specificity relative to need.
- Timeliness: the availability of information relative to the time it is needed.
- Trust: the credibility of the information source.
- Confidence: the willingness to use the information.

(In this context, *information* is needed for assigned tasks at the individual level, for missions at the mission level, and for the spectrum of missions at the enterprise level.)

For example, timeliness measures the receipt of information relative to when it is needed. At the individual level, the attributes *relevance* and *completeness* use for their denominators the universe of information considered to be relevant for the individual's assigned tasks. At the mission or enterprise level, the denominator would be the universe of information considered to be relevant for the conduct of the mission or the enterprise respectively.

A note of caution! *Relevance* is a concept that is well-understood, but it is very, very difficult to apply in practice. With the exception of well-understood and predictable domains, it is extremely difficult, if not impossible, to enumerate in advance all of the information that may be relevant in a given situation or mission. Afterwards, it is possible to identify information that was felt to be needed but was not available at all or not available to the individual or organization that needed it. Many traditional system designers, perhaps the same ones that clamor for a freeze on requirements, rely on information exchange requirements (IERs) to determine how to design and build a system. This practice is now widely understood to be flawed because it is impossible to know the information that will be needed in advance, particularly for the kinds of situations and missions that we will face in the 21st century. Thus, we recommend that the relevance-related measures *relevance* and *completeness* be used after the fact to indicate the agility of the information processes provided.

Information reach

The second component of information position is information reach. This set of metrics applies only to the mission and enterprise levels. These variables reflect the distribution of information over individuals and/or organizations. We have identified two accessibility measures and two shared information measures. Each pair contains one measure that is based on all of the available information and one based on only the portion of the available information that is relevant.

- Accessibility Index (all): the proportion of the available information that is accessible.
- Accessibility Index (relevant): the proportion of the relevant available information that is accessible.
- Index of Shared Information (all): the available information that is accessible by two or more members (COI, mission, or enterprise perspective).
- Index of Shared Information (relevant): the available and relevant information that is accessible by two or more members (COI, mission, or enterprise).

First, there is the issue of how much of the available information can be accessed by any given individual (the accessibility index). In reality, this amount will differ from individual to individual and from organization to organization. Measures of interest will include both the mean and the variance of the distribution across the mission or enterprise. The basis for these calculations could be either *all available information* or *all available relevant information*. The former would clearly be a better reflection of the organization's information agility, while the latter would be a better reflection of situational utility.

One way to make this measure more meaningful would be to calculate it on a community of interest basis. However, it should be remembered that individuals and organizations are members of multiple COIs. Whatever approach one selects to measure information reach, this result is not as meaningful if the measure is viewed in isolation.

Information reach becomes more meaningful when viewed along with other information quality-related metrics. If, for example, a *successful* mission is rated with a relatively low score for the distribution of information (based on all available information) while receiving a high score for information completeness, then the completeness metric is likely to have been of more importance for that mission because the "inaccessibility" factor did not prevent mission success. The lack of access or availability (a lower score on reach) did not prove to be substantially detrimental in this case, at least at the mission level. At the enterprise level, it becomes more complicated because we need to view this from an agility perspective where we do not know what will be needed in future operations. If, on the other hand, a mission or enterprise receives a high score on information distribution but a low score on relevance or completeness, then, knowing both, one can use this as a point of departure to improve information quality and therefore achieve greater effectiveness and agility. Thus, the use of information reach in two different formulations and with other information-related measures can serve to enhance our understanding of an organization or task, as well as serve as a diagnostic tool.

The accessibility index discussed above measures how widespread the access to the information is. Another measure that is of interest is the extent to which information is shared, that is, how much of the information (access) is common to two or more organizations. As access (the accessibility index) approaches 1, the set of information that any *n* parties have access to will also converge on 1. That is, everyone will have access to the same information. When the accessibility index is, as is the case today, far less than 1, it is important to know how much information members of COIs have in common. If information processes are designed well, the index of shared information will rise more rapidly than the accessibility index.

Information security

In addition to the richness of the information and its availability across mission participants or across the enterprise, the value of information depends on the degree to which it is secure. Information assurance has always been important, but it is arguably more important as our concepts of operation and C2 Approaches increasingly rely on improved situation awareness and the ability to interact with one another dynamically. Information assurance is required to generate the trust and confidence in the information that is necessary for these strategies to be effective. Information assurance is about protecting our information and information sources from a variety of attacks that can result in denying information to some or all participants and/or compromising the information itself. Specifically, the security of our information is a reflection of five key attributes: privacy, integrity, authenticity, availability, and non-repudiation.[95]

Information interactions

The conceptual model explicitly includes measures of the quality of interactions that take place in both the information and social domains. It is these interactions that play a major role in determining the quality of the information position and

[95] Information assurance is defined as information operations that protect and defend information and information systems by ensuring their availability, integrity, authentication, confidentiality, and non-repudiation. This includes providing for restoration of information systems by incorporating protection, detection, and reaction capabilities. DoD Dictionary of Military and Associated Terms.

the ability of mission participants to make sense of the situation and take action in a synchronized manner. Information interactions may differ from one another in a number of dimensions. They can differ in the form of information that is involved and the nature of the interaction itself. Figure 18 presents a number of different forms that information can take and the attributes of an interaction.

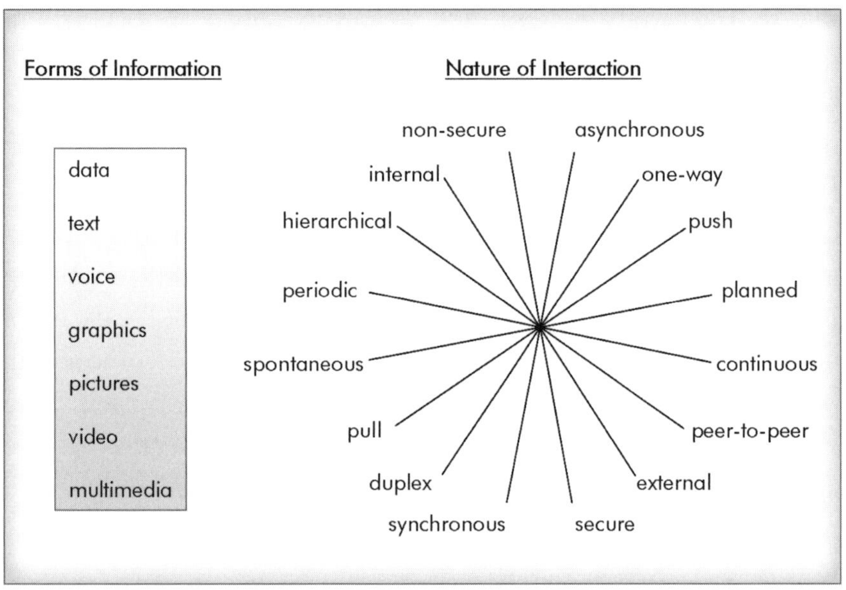

FIGURE 18. INFORMATION INTERACTIONS

Different forms of information affect the amount of information that can be conveyed in a given amount of time, as well as the degree and rate of its absorption by individuals. The nature of the interactions will affect the propagation of information across the participants.

Quality of information position

Relationships

The four components of information position (information richness, information reach, information security, and information interactions) are interrelated and are affected by, among other things, the quality of ISR (intelligence, surveillance, and reconnaissance), transport, and information services. Figure 19 depicts these relationships.

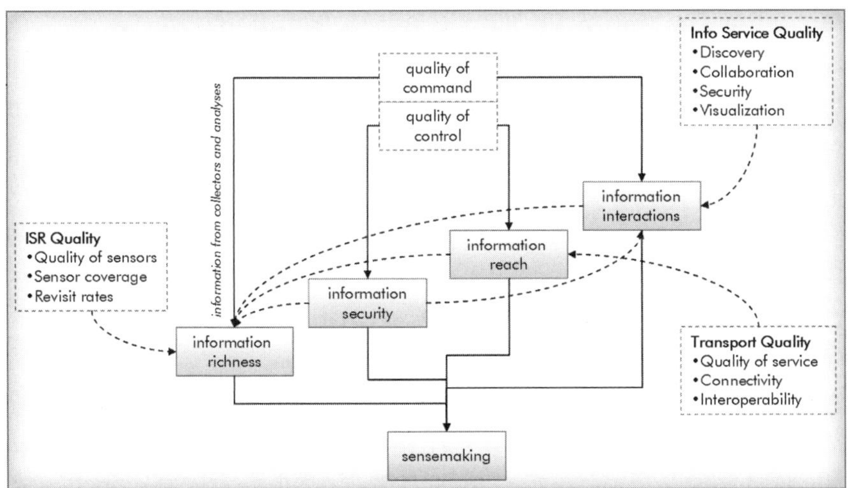

FIGURE 19. INFORMATION POSITION COMPONENTS AND RELATIONSHIPS

Information richness is, in the short run, limited by the nature of the information that exists, the information that is available or can be collected, and of course, what is known in advance. Thus, information richness depends to a significant degree on the quality of ISR. Several attributes of information richness depend on the age or currency of the information, which in turn depend on the quality of transport capabilities that may facilitate or constrain information flows. Attributes of information reach also depend on where information flows, which determines the resulting topology of information. Information flows depend on more than the capabilities of the transport layer; they also are a function of the nature of the information-related interactions that take place among individuals, organizations, and systems. These in turn are affected by the quality of information services provided. In many organizations, some information and information exchanges are restricted to authorized individuals. Thus, an integral part of information services is related to information assurance that will directly affect what and how information is disseminated and accessed, as well as the trust that individuals and organizations have in information. Information richness depends on information reach, information interactions, and information security. These in turn are influenced by the quality of the network (both the transport and service layers) as well as the conditions that govern information processes and interactions that are established by **Command and Control**.

Although it is not customary to think about it in this way, **C2** also directly affects all of the components of the quality of the information position. Command has a direct effect on information richness by the nature of the assignments that are made and how they are distributed. If a good match is achieved between task assignments and information accessibility, increased information richness will result. Command also sets the conditions that influence the nature of the interactions that will take place between and among individuals and organizations, and establishes security policies. These command decisions, by affecting information security and interactions, affect information richness and reach.

In the long run, investments in any number of things can affect one's information position. Investments in information gather-

Quality of information position

ing can make more information or better quality information available. Investments can be made to increase the quality or quantity of a range of information sources including the organic sensors of participants (what are known as technical means) and access to and relationships with open sources. Increasingly, who we develop relationships with affects what information we have readily available and the degree to which information sources are understood and trusted. Investments that will affect information richness also include not only the development of new sensors or ways of positioning them (e.g., UAVs) but also investments in the ability to turn data into information, such as improved language skills and training. As DoD implements its post-and-smart-pull policy, increased investments in metadata tagging, discovery tools, and training will be needed to help individuals and organizations shape their own information positions.

Security policy and the means to carry it out are not independent considerations, but need to be understood as they affect the key components of information position. Information position, as defined here, permits the full range of information-related investments to be put into proper context. This is of particular significance for information assurance because it has historically not been fully integrated into the decisionmaking processes that address information-related capabilities. With information richness having information assurance-related attributes (e.g., trust and confidence) and influenced attributes (e.g., completeness, accessibility index, and extent of shared information) that exert opposing influences, trade-offs are easier to understand.

In the final analysis, the quality of an information position at the individual, mission, and enterprise levels is a complex concept that involves a large number of dimensions, variables, and relationships among them. The main determinant of the quality of an information position—who has access to what—will create the conditions that shape sensemaking processes. Information is a currency, something individuals and organizations trade and use to assess relationships as well as the situation. The quality of an information position will also assist in determining what **Command and Control** Approaches make sense for the situation and/or organization.

Quality of Sensemaking

Sensemaking takes place in the cognitive and social domains.[96] Awareness, knowledge, and understanding are states that reflect the quality of sensemaking, while decisions are the products of sensemaking. These, like an information position, need to be examined at the individual, mission, and enterprise levels. Sensemaking is an ongoing activity, but one that can be viewed in the context of a particular situation. This discussion of sensemaking will start with individual sensemaking and then proceed to team sensemaking.

---

[96] Sensemaking: The process by which individuals (or organizations) create an understanding so that they can act in a principled and informed manner. When a sensemaking task is difficult, sensemakers usually employ external representations to store the information for repeated manipulation and visualization. Working can take different forms such as logical, metaphorical, physical. ISTL-PARC. <www2.parc.com/istl/groups/hdi/sensemaking/gloss-frame.htm> Jan 2006.

Individual sensemaking

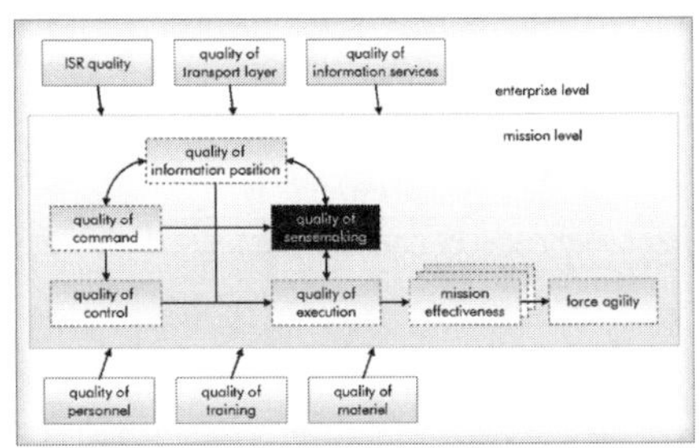

Over time, individuals develop mental models that determine how they perceive and understand information. They also invariably develop perceptual filters and biases that affect how they interpret and understand information, as well as how they interact with others and how they behave in specific situations. The cognitive state of an individual at any point in time also affects their ability to process information. Stress and physical conditions contribute to this state. Education, training, and experience play an important role in determining how efficiently and effectively individuals perform this function in a variety of circumstances.

Whether an individual trusts or has confidence in a particular piece of information is influenced by perceptions of the source, the security of the information system, and by other *a priori* perceptions and understandings that all impact the sensemaking process. Thus, the ability of individuals to draw on whatever reservoir or libraries of pertinent knowledge (sometimes called task knowledge) and understanding they have developed over time depends on the MCP elements of personnel, education, and training. These factors set the conditions that affect the relationships between the quality of information and the quality of awareness, understanding, and decisions. Figure 20 takes a value view of individual sensemaking and identifies key factors that influence this part of the value chain.

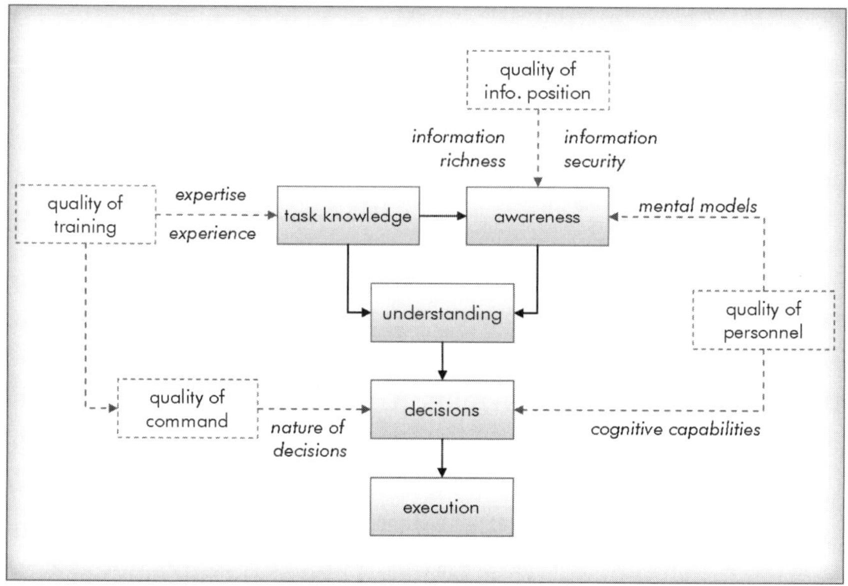

FIGURE 20. INDIVIDUAL SENSEMAKING

Collective sensemaking

Collective sensemaking involves a group of individuals that collaborate in some fashion on a task or mission. This group may be part of an established organization with doctrinal practices and processes, or they may be ad hoc with or without established practices and processes. The characteristics and nature of the group and the characteristics of the individual members, taken together, create a set of conditions that affect how sensemaking is accomplished and, from the value perspective, the quality of sensemaking.

The concepts central to individual sensemaking—awareness, knowledge, and understanding—also have group or team instantiations. In addition to these concepts, an understanding of collective sensemaking requires that we consider, as measures of value, shared awareness, shared knowledge, and

Quality of sensemaking

shared understanding. We also need to characterize and understand the interactions that take place among individuals and groups of individuals during the process of sensemaking. In the initial book on NCW (*Network Centric Warfare*, 1999), the discussion of the cognitive domain was limited to awareness, while the discussion of the social domain was limited to shared awareness, collaboration, and self-synchronization. This simplified discussion was meant to avoid complicating what have become known as the tenets of NCW.

Understanding Information Age Warfare (2001) discussed the need to fuse knowledge with information to develop battlespace awareness to make assessments of the situation. It also discussed sensemaking and identified *understanding* as a part of sensemaking,[97] and it expanded the NCW discussion of collaboration[98] by identifying the dimensions of collaboration and what would constitute "maximum" collaboration.[99] Figure 21 incorporates all of these concepts into a value view of collective sensemaking.

As can be seen from Figure 21, many factors influence how sensemaking is accomplished, as well as its quality. First, command determines what roles and responsibilities individuals and organizations take on and hence affects the nature of the cognitive tasks to be performed. Furthermore, command, by influencing the *information position*, contributes to influencing what information-related resources are available and how they are distributed. Finally, command, by policy and doctrine, determines, encourages, and/or constrains the interactions that take place. The characteristics of the personnel involved

---

[97] Alberts et al., *Understanding Information Age Warfare.* pp. 136-145.
[98] Ibid., pp. 185-202.
[99] Ibid., pp. 200-202.

affect individual cognitive performance and influence the interactions that take place. *Training and education* are second only to command in influencing how well cognitive tasks are performed and in determining the nature of the social interactions that take place.

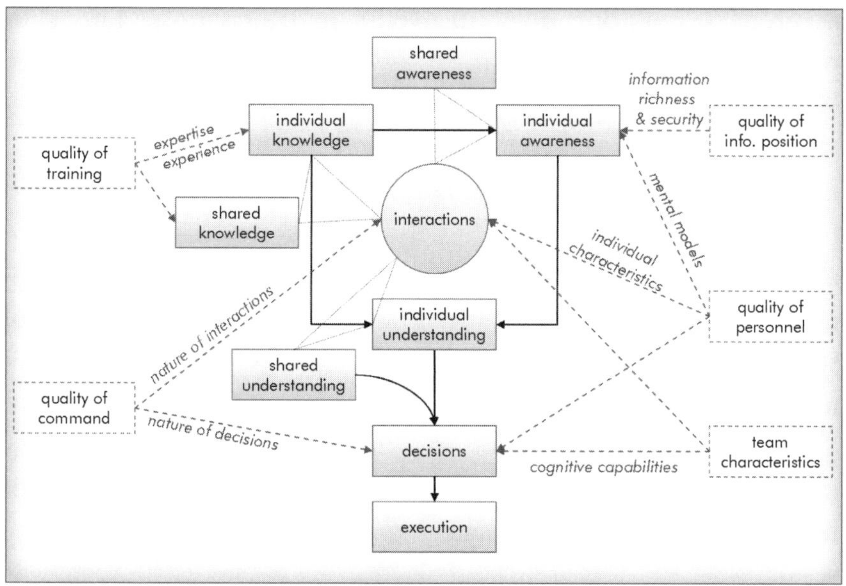

FIGURE 21. COLLECTIVE SENSEMAKING

Team behaviors and capabilities are shaped by *team characteristics* as well as *individual characteristics.* These have a great deal of influence on how individuals and teams perceive reality and on how individual team members relate to one another. Culture plays a major role in shaping perceptions, sensemaking approaches, and decisions. In the final analysis, the option space that an individual or a team considers may be the most important determinant of the nature of the decision. Not only is the option space shaped by culture, but the value that is associated with each particular outcome is, in large part, determined by cultural values.

Quality of sensemaking

A great deal of discussion has focused on the problem of "runway teams," teams that meet for the first time as they deploy. It has been repeatedly learned that these ad hoc teams take time to harden into effective organizations. This need for a period of hardening persists despite the fact that as individuals they may have been well-trained and experienced in their assigned roles. Given that most missions will include a set of participants from a variety of organizations, it is more likely that a significant portion of the team will not have worked together before. Thus, it is important to better understand team dynamics and how teams composed of members from a variety of organizations with a variety of skills and experiences can come together into an effective unit more quickly. Understanding team dynamics and the effects that individual and team characteristics, organization and doctrine, and **Command and Control** Approaches play will require the kind of conceptual model that has been presented here.

In Figure 21, the box *decisions* was not labeled "shared decisions" because, in undertakings with the complexity in which we are interested, a great number of decisions will be made both individually and collectively. Understanding this final phase of the sensemaking process will, among other things, require us to understand the nature of decisions that are (1) individually made, (2) individually made but collectively influenced, as well as those that are indeed (3) collective decisions.

Sensemaking metrics

There are a number of MoMs associated with sensemaking. These include the quality of *individual* awareness, knowledge, and understanding as well as the quality of *shared* awareness, knowledge, and understanding. They also include the quality of interactions and the quality of decisions.

Awareness or situation awareness is an individual's perception of the information about the situation. The term *awareness* was introduced into the discussion of **C2** to create a bright line between the information that is available or the information displayed in a "common operational picture" and what an individual perceives regarding the situation. Traditionally, analyses of **C2** systems used the quality of information as a surrogate measure for the quality of awareness. It was not until there was a desire to closely examine the tenets of NCW that the need for measuring awareness directly was more widely recognized. Awareness is the cognitive associate of information and the metrics for the quality of awareness mirror those associated with the quality of information.

In previous discussions of awareness, we presented a reference model of the situation (see Figure 22).[100]

This figure identifies the components that, taken together, define situation awareness or what can be known about a situation. What needs to be known is situation dependent and is a subset, usually a rather small subset, of "total" situation awareness. Awareness consists of an amalgam of *a priori* knowledge and current information, which include the:

- Capabilities, intentions, and values[101] of the various players (blue, red, others);
- Nature of the task(s) at hand (intent, constraints);

---

[100] Alberts et al., *Understanding Information Age Warfare.* p. 121.
[101] Values here refer to cultural norms, customs, and priorities.

- Relevant characteristics of the environment; and
- Previous and current status/states (location, health) of the entities of interest.

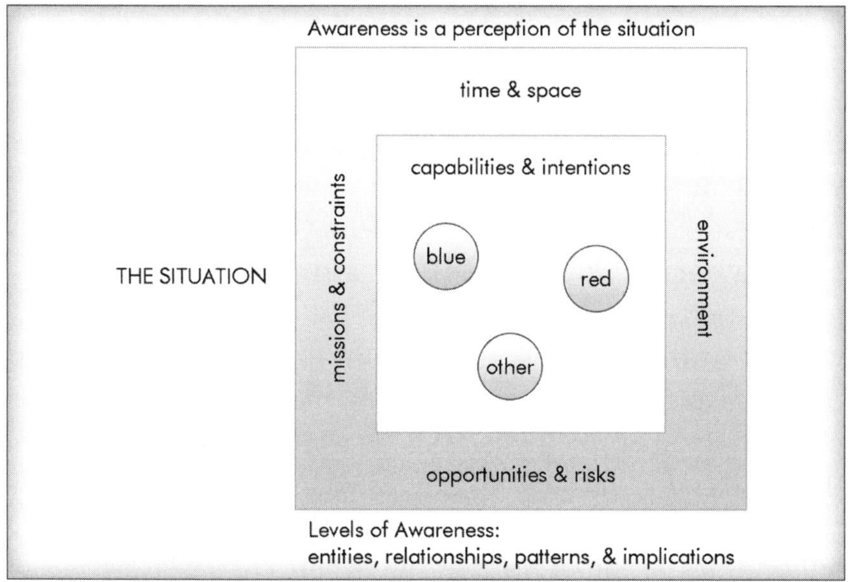

FIGURE 22. REFERENCE MODEL: THE SITUATION

What needs to be known depends on a combination of the nature of the decisions that need to be made and the circumstances that obtain. The recognition that a decision is needed, a kind of decision in and of itself, may not require very much to be known, certainly less than is required for the decision itself. Different kinds of decisions require different types of knowledge or degrees of awareness. A simple decision, one that has a pre-existing template that maps the value of one or more variables to one or more specific actions, requires only awareness of the values[102] for the specific set of variables included in the template.

[102] *Value* refers to the value of a variable (e.g., 108 = number of tanks in the area).

In almost all cases, an essential ingredient of awareness is *intent*, for this determines both when a decision needs to be made and the decision template. In the language of analysis, this is the formulation of the problem.[103]

This reference model also identifies four levels of awareness: entities, relationships, patterns, and implications. An example that is often used to explain the way experts think involves an experiment with a chess board.[104] When both novice and non-chess players are shown a game in progress, they can only correctly recreate the positions for a small number of pieces, while experts can usually correctly recreate the entire board. A conclusion that could be drawn from this result is that expert chess players have extraordinary memories. However, when chess pieces are placed randomly on a chess board, it turns out that neither the novices nor the experts can correctly replicate their positions. Thus, the conclusion about memory does not hold up. Rather it seems that the expert chess players recognized the patterns they saw in an actual game and were able to reconstruct the positions of pieces from an understanding of these patterns.

These results were found in an experiment[105] conducted to test various visualizations of the battlefield. In this experiment, it was found that no one was able to recall more than about

---

[103] See: NATO *Code of Best Practice.* pp. 37, 53-67.

[104] As demonstrated in:

Chase, W.G. and H.A. Simon. "Perception in Chess." *Cognitive Psychology*. No 4. 1973. pp. 55-81.

Saariluoma, P. "Chess Players Recall of Auditory Presented Chess Positions." *European Journal of Cognitive Psychology*. No 1. 1989. pp. 309-320.

[105] "SCUDHunt and Shared Situation Awareness." Experiment conducted by ThoughtLink and the Center for Naval Analyses for the Defense Advanced Research Projects Agency (DARPA).

one third of the entities displayed. Yet some were able to understand the situation by recognizing patterns and drawing inferences from these patterns. Measuring awareness requires that the subjects' awareness of entities, relationships, patterns, and inferences all be recorded. This will provide a basis for understanding the state of the cognitive chain and thus a basis for understanding what the weak link is and potential ways to improve the result.

Usually there is information on only a fraction of the entities of interest at any given point in time. However, knowledgeable individuals can recognize patterns in the available information and, consciously or unconsciously, draw inferences from these patterns that serve to fill in the blanks or connect the dots. Generally speaking, the more experienced or expert one is, the greater their ability to recognize and understand patterns becomes, and hence the better able they are to fill in the blanks. Missing information may, in fact, be more important than the available information because it may mean a deviation from a pattern, faulty information, or perhaps deception. A common cause of error is filling in the blanks unconsciously or without proper attention and care. This is referred to as *premature narrowing of vision*. In common language, it is referred to as jumping to conclusions.[106]

Among the implications that can be drawn from entity and pattern information are both the opportunities and risks present in the situation. These require significant experience and expertise. Experienced individuals also pay attention to what is missing from the picture. Among the risks is the identi-

---

[106] For an in-depth discussion of risk, bias, and decisionmaking, see: Kahneman, Daniel and Amos Tversky. "Prospect Theory: An Analysis of Decision under Risk." *Econometrica*. Vol 47, Iss 2. March 1979. pp. 263-292.

fication of residual uncertainties, the important things that are not known. These may include how someone will react to a given action or the effects of some action.

Awareness is always informed by both contemporaneous information and *a priori* knowledge. The capabilities of various players are usually known in advance, but these capabilities will be affected in real time and thus as a situation progresses, awareness depends more on information and less on *a priori* knowledge. Often relationships are known in advance and, when combined with real-time information, allow patterns to be identified from fragmentary information.

The attributes of the quality of awareness are provided below in two groups.[107]

Individual awareness, situation *independent* attributes:

- Correctness: the extent to which awareness is consistent with ground truth.
- Consistency: the extent to which awareness is consistent with prior awareness.
- Currency: the time lag between the situation and awareness of it.
- Precision: the degree of refinement, level of granularity, or extent of detail.

[107] These attributes, like others in this section, are modified from those developed for the NCO Conceptual Framework.

Quality of sensemaking

Individual awareness, situation *dependent* attributes:

- Relevance: the proportion of the awareness that is related to the task at hand.
- Completeness: the degree to which awareness is sufficient to achieve understanding.
- Accuracy: the precision relative to need.
- Timeliness: the awareness attained relative to the time it is needed.
- Confidence: the willingness to draw conclusions based on awareness.

Awareness and knowledge are the inputs to a process of understanding. Understanding a situation requires that one be able to predict how events are likely to unfold over time, including the characterization of alternative futures and the identification of residual uncertainties. Understandings form the basis for decisions, the final stage of sensemaking. The attributes associated with individual understanding parallel the attributes associated with information and with awareness.

Individual understanding, situation *independent* attributes:

- Correctness: the extent to which understanding is consistent with ground truth.
- Consistency: the extent to which understanding is consistent with prior understandings.
- Currency: the time lag between the situation and understanding it.
- Precision: the degree of refinement, level of granularity, or extent of detail.

Individual understanding, situation *dependent* attributes:

- Relevance: the proportion of the understanding that is related to the task at hand.
- Completeness: the degree to which understanding is sufficient for decision(s).
- Accuracy: the precision relative to need.
- Timeliness: the understanding achieved relative to the time it is needed (decision).
- Confidence: the willingness to decide based on understanding.

The final step in the process of sensemaking is the decision that is made. All decisions are not created equal. There are simple decisions that consist of one-to-one mappings from specific understandings to specific decisions (options or alternatives). That is, an accepted decision rule exists that fits the circumstances. In the case of simple decisions, the available options are known. Also known are the relative values of the options for different circumstances and hence the best option as a function of circumstances. Complex decisions require that suitable options be generated as well as assessed.[108]

There are two basic approaches for determining the quality of a decision. In short, these equate to: "Was this the best decision, given the information available?" or "Was this the right decision made at the right time, given what happened?" The discussion regarding the appropriate approach to measuring the quality of decisions is essentially the same discussion we had earlier about the use of mission effectiveness to measure

---

[108] These types of **Command and Control** decisions are discussed in: Alberts et al., *Understanding Information Age Warfare.* pp. 152-154.

the quality of **Command and Control**. For these reasons, the quality of decisions will be measured using a set of attributes that do not contain the attribute "correctness." Rather, the attributes we suggest for measuring the quality of the decisions are those that one would want to know to answer the first of the two questions.

Decisions, situation *independent* attributes:

- Consistency: the extent to which decisions are consistent with prior decisions.
- Currency: the time lag between situation awareness and making a decision.
- Precision: the degree of refinement, level of granularity, or extent of detail.

Decisions, situation *dependent* attributes:

- Appropriateness: the extent to which decisions are consistent with existing understanding, command intent, and values relevant to the situation.
- Completeness: the extent to which relevant decisions encompass the necessary:
  - Depth – range of actions and contingencies included.
  - Breadth – range of force elements included.
  - Time – range of time horizons included.
- Accuracy: the appropriateness of precision of the decision (plan, directives) for a particular use.
- Timeliness: the time of the decision made to the time it is needed for action.
- Confidence: the willingness to act.

All things being equal, agile decisions (those that work in the face of changes in circumstances) are preferred to decisions that are brittle and will only work well if the situation is as understood and anticipated. However, at times it is better to make a series of less agile decisions and remake them if and when changes are required. Agility can also be created by making decisions that increase the number and variety of available options, but option creation is never a goal in itself and must be coupled with decisions to act effectively.

Agile **C2** is a function of both the agility of decisionmaking and the agility of the decisions made. Given the nature of 21st century missions, it would be hard to over-emphasize the importance of **C2** agility. In general, the more familiar the situation is, the better it is understood, and the greater its predictability becomes, and so there is less need for agility and one can optimize more and take the risks associated with a brittle decision.

To a considerable extent, cultural norms play a major role in the nature of the decisions made and the willingness of individuals and organizations to change course.

These are the attributes of agility as applied to decisions:

- Robustness: the degree to which a decision is effective across a range of situations.
- Resilience: the degree to which a decision remains applicable under degraded conditions or permits recovery from setbacks.
- Flexibility: the degree to which a decision allows force entities to maintain flexibility (i.e., incorporates multiple ways of succeeding).

Quality of sensemaking

- Adaptability: the degree to which a decision facilitates force entities' ability to alter or modify the decision or decision process.

In addition to these value-related attributes, it is best practice to observe or solicit the nature of the decision approach utilized. For example, was the decisionmaking process a naturalistic one? Was the option chosen because it represented the best expected outcome or utility, or was the selection made to minimize risk?

Shared awareness plays a major role in the tenets of NCW, serving as the pivot point for new **C2** Approaches. The distinctions between awareness, knowledge, and understanding provide a greater degree of granularity and hence, they provide an opportunity to better diagnose problems and opportunities to improve sensemaking. Shared awareness (A), knowledge (K), and understanding (U) are all measured in a similar fashion.

Situation *independent* attributes:

- Extent: the proportion of A, K, U that is the same across the population of interest (COI, mission, enterprise).
- Correctness: the extent to which shared A, K, U is consistent with reality.
- Consistency: the extent to which shared A, K, U is consistent with prior shared A, K, U.
- Currency: the time lag between the situation and shared A, K, U.
- Precision: the degree of refinement, level of granularity, or extent of detail.

Situation *dependent* attributes:

- Relevance: the proportion of shared A, K, U that is related to the task at hand.
- Completeness: the degree to which shared A, K, U is sufficient.
- Accuracy: the precision of shared A, K, U relative to need.
- Confidence: the willingness to draw conclusions or take action based on shared A, K, U.

The extent of shared awareness, knowledge, and understanding can be measured relative to different baselines: teams, COIs, mission participants, or enterprises. Each of these sets of measures of extent is useful for some purpose, although not all of them may not be applicable to a given analysis. When one is considering enterprise capabilities or any investment that can serve multiple purposes and missions, it is important to use the broadest definition of the population because the nature of current missions makes it impossible to know exactly all of the information, individuals, and organizations with which it may be necessary to exchange or interact.

The preceding attributes and definitions address the measures of all of the sensemaking concepts depicted in Figure 21 with the exception of *interactions*. Interactions among individuals and/or organizations are important not only for sensemaking but also for information-related transactions and tasks and for execution. The nature of interactions and the attributes related to the quality of interactions are addressed in the next section.

Quality of sensemaking

Quality of interactions

Understanding interactions is an important key to understanding **Command and Control**. The nature of the interactions that take place in the information, cognitive, social, and physical domains greatly influences the quality of information, sensemaking, and actions. Figure 18 presents a number of different forms of information and identifies attributes of information-related interactions. Many of these attributes apply to the cognitive and social domains as well, for example, the temporal characteristics of an interaction, whether it is periodic or continuous.

The word *interaction* is quite general and can refer to a wide variety of behaviors. For the purpose of exploring and understanding **C2**, we are interested in a set of these behaviors that involve a common or shared purpose or objective. Collaboration is defined as "working together for a common purpose."[109] There is a spectrum of collaboration. Words like sharing, coordination, consultation, synchronization, and integration could be used to describe differing degrees of collaboration. But comparing what one may mean by coordination versus consultation is difficult if not impossible unless we explicitly define the dimensionality of collaboration. Once we define the dimensionality of collaboration, we can fix the point on each dimension that corresponds to a given descriptor (e.g., coordination). In fact, it is typical to think about coordination in varying degrees (e.g., close coordination).

When looking at a spectrum, it is often helpful to define the endpoints, in this case maximum collaboration. Maximum col-

---

[109] Alberts et al., *Understanding Information Age Warfare*. pp. 185-203.

laboration, as defined below, corresponds to an endpoint on each one of the dimensions of collaboration.

The maximums are:

- Extent: *Inclusive* – all participants involved (collaboration cuts across organizational, functional, spatial, and temporal boundaries including echelons of command).
- Access: *Full and equal access* – all participants have equal access to all other participants.
- Communications: *Unconstrained* – sufficient bandwidth.
- Level of participation: *Participatory* – all participants fully engaged.
- Frequency: *Continuous* – participants engaged without interruption.
- Synchronicity: *Synchronous*.
- Richness: *Rich* – multimedia, face-to-face.
- Scope: *Complete* – involves data, information, knowledge, understanding, decisions, and actions.

Minimum collaboration would (1) include a small number of participants from a single organization, echelon, or function; (2) grant limited access to most participants; (3) have significantly bandwidth-constrained interactions; (4) include only a few active participants, most being passive; (5) occur infrequently; (6) be asynchronous; (7) be limited to voice or email; and (8) involve only the exchange of data. Such minimum collaboration would be unlikely to generate results different from those generated by an individual working alone.

Quality of interactions

QUALITY OF COMMAND

The *quality of command* box in Figure 16 refers to the quality of mission command. As with the quality of decisions, some are tempted to measure the quality of mission command by looking at the accomplishment of the mission. For the same reasons that this is not a good idea for the quality of decisions, it is not a good idea for the quality of command. A more direct approach is preferred, one that focuses on the degree to which the functions of command are accomplished.

As delineated in Chapter 4, the following functions are associated with **C2** (or management) of a given undertaking:

- Establishing intent;
- Determining roles, responsibilities, and relationships;
- Establishing rules and constraints; and
- Monitoring and assessing the situation and progress.

Thus, the quality of mission command is related to the accomplishment of these four functions and is a function of the quality of intent, the quality of the information position, the quality of interactions, the quality of sensemaking, and the quality of execution. These quality measures map to the four functions of mission command in the following manner. The function of *establishing intent* maps into a measure of the quality of intent. The command functions of *determining roles, responsibilities, and relationships* and *establishing rules and constraints* are meant to create an effective organization by shaping the processes and activities associated with information, sensemaking, and execution. The assignment of roles and responsibilities also shapes the way the function of control is performed. Thus, these command functions are best assessed in terms of the

Quality of command

qualities of information position, sensemaking, execution, and control. The degree to which the command function *monitoring and assessing the situation and progress* is accomplished can be measured by the qualities of information position and sensemaking. Figure 23 depicts the mappings from command functions to the measures of quality of command.

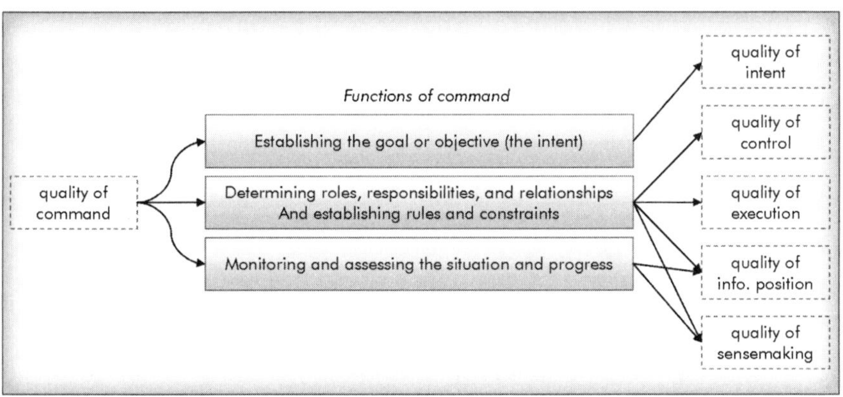

FIGURE 23. QUALITY OF COMMAND

All of these quality measures, with the exception of the *quality of intent*, have been previously discussed in this chapter. The quality of intent needs to factor in a consideration of the (1) existence of intent, (2) the quality of its expression, (3) the degree to which participants understand and share intent, and in some cases, (4) the congruence of intent.

QUALITY OF CONTROL

Command establishes intent and creates the conditions under which the function of *control* is performed. The function of control is to determine whether any adjustments are required to intent, its expression, or to the established roles, responsibilities, and relationships, and the rules and constraints that are in effect. The *quality of control* reflects the accomplishment of the functions associated with control. The first function of control is to recognize that there is an "occasion for a decision,"[110] that is, that something is not going as anticipated, expected, or planned. This is, in effect, the fourth command function: monitor and assess the situation and progress. This function is traditionally delegated after the nature of the employment of the force and the other functions of command have been discharged to instantiate a mission capability package. When it is recognized that some adjustments are required, the function of control is to, within the limits established by command, make changes to the established roles, responsibilities, and relationships, and the rules and constraints that are in effect. If it is determined that the changes that are needed go beyond the authority or means that have been delegated, then the function of control is to inform command of the nature of the decisions that need to be made.

The function of control is essentially a decisionmaking function accomplished within the parameters set by command. Thus, the attributes that define the quality of control are the same attributes that define the quality of decisions.

Quality of Execution

Decisions lead to actions[111] that, in turn, create a set of effects that are designed to alter a situation in a manner consistent with

---

[110] Occasion for a decision: in legal parlance, this refers to the advent of a situation in which it becomes possible or necessary for a court to rule upon a law and thus set a precedent, enforce an existing law, or declare a law to be unconstitutional.
[111] A decision may be made to delay a decision, to seek more information, or to do nothing at that point in time. These decisions also result in actions: the act of seeking information or the act of waiting or inaction.

intent. While the term *fog of war* refers to a lack of understanding and/or shared understanding in the midst of a conflict, the term *friction of war* refers to less than perfect execution.

Separating decisions and execution is quite arbitrary. The traditional literature on **Command and Control** does not appropriately recognize the myriad decisions that are made across the force, choosing to focus only on command decisions and more often than not focusing on one commander or headquarters. However, it is recognized that there are command decisions at all levels of command. The problem with this way of thinking is that command decisions are defined, not in terms of the functions of command, but in terms of who makes the decision. This is simply inappropriate if one is trying to sample the full range of approaches to accomplishing the functions associated with command (and control). Thus, the traditional approach bundles the vast majority of the decisions that are made over the course of an operation with execution. Execution, unlike a particular decision that can be considered to be made at a given point in time, takes place over some period of time. During this period, many decisions will be made that influence the quality of execution because these decisions will determine what actions will be taken, when they will be taken, and how they will be taken.

This conceptual model, designed as it is to facilitate exploration of a variety of approaches to command and to control, separates decisionmaking from the execution of decisions. It takes the position that the quality of decisions should be measured by the same criteria regardless of the individual or organization making the decision or the type of decision it is.

Quality of execution

Figure 24 looks at the factors that influence the quality of execution as well as the three components of execution quality. The quality of execution is a function of how well individual tasks (those that flow from decisions) are performed, how well these individual actions are synchronized, and the agility associated with execution. The quality of the execution of a specific action or task is a function of the competence, expertise, and experience of the individuals or organizations that are involved, the quality of their information position, and the execution agility of the force (individual or team).

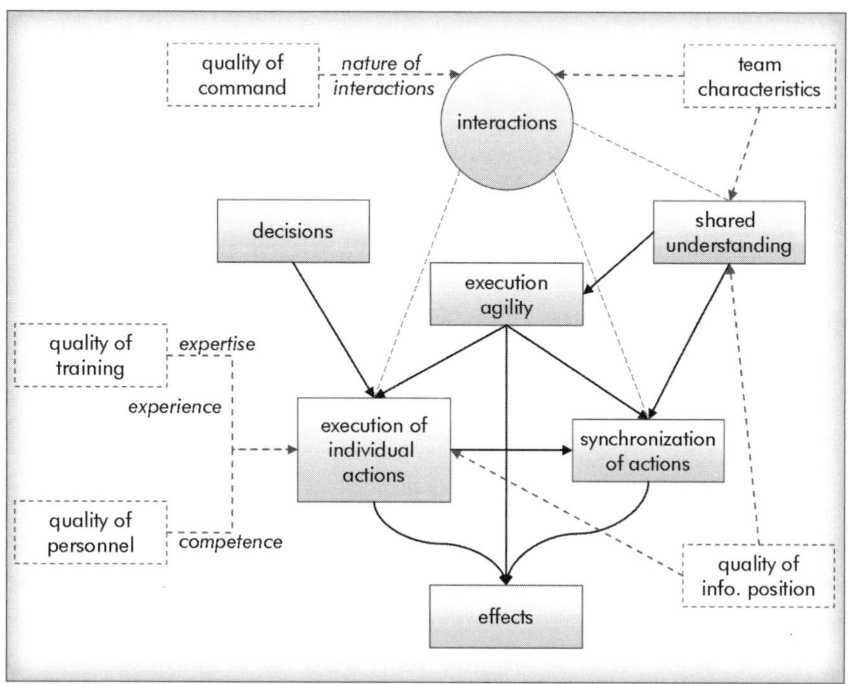

FIGURE 24. QUALITY OF EXECUTION

The degree to which these actions are synchronized is a function of shared understanding, the nature of the interactions that take place, and how well individual tasks are performed.

Synchronization is an arrangement in time and space, where the "space" can no longer be interpreted as the physical battlespace, but must be interpreted as the "effects space." The effects space includes all of the domains: the information, cognitive, and social in addition to the physical. The expansion to the effects space corresponds to the move from a traditional objective of massing the force to the *massing of effects*.

Synchronization can come about in a variety of ways, *self*-synchronization being just one of them.[112] Measuring the degree of synchronization (of actions) involves categorizing the relationships between and among the actions taken. In the most basic terms, two or more actions are either (1) unrelated, (2) in conflict with one another, or (3) re-enforcing one another (synergistic). Traditional C2 Approaches focused on achieving deconfliction as opposed to achieving synergy. NCW and Power to the Edge concepts are focused on achieving synergy. Measuring the degree of synchronization involves developing a function that integrates (in the sense of a calculus) across the actions, relationships, and time.[113]

Execution agility is the third component of the quality of execution. All six of the components of agility[114] can come into play. Under most circumstances, it is easier to observe the execution of decisions than the decisions themselves, and thus under most circumstances the quality of execution will be used as a surrogate for the quality of decisions. In experiments of

---

[112] Self-synchronization requires that there be sufficient levels of shared awareness, training, and understanding of not only the situation but also of the capabilities and behaviors of appropriate levels of the group.
[113] An in-depth discussion and examples can be found in:
Alberts et al., *Understanding Information Age Warfare*. pp. 205-237.
[114] Further elaboration on the six attributes can be found in:
Alberts and Hayes, *Power to the Edge*. pp. 123-163.

various kinds and in some lessons learned activities, attempts can be made to instrument decisions and decision processes. Eventually the nature of the link between the quality of decisions and the quality of execution can be established.

Force Agility

Agility is the appropriate metric to use for Information Age organizations. NCW and Power to the Edge principles highlight the importance of the quality of information position, quality of sensemaking, quality of command, quality of control, and quality of execution. These variables[115] constitute one side of an equation, a set of the independent variables with *force agility* as the dependent variable. They form a scenario-independent value chain. These quality metrics are directly related to the quality of each of the elements of a mission capability package and the results of efforts to shape the security environment in which the force operates.

Informing the Value View

The value view presented in this chapter is a vessel, one that needs to be filled with both parametric relationships and value ranges. It is one thing to state the quality of one's information position as a function of this or that, and quite another to specify the form of the equations that link the quality of information position to its determinants.

---

[115] These quality measures are multi-dimensional. The NATO research group SAS-050 refers to these as concepts and notes that they are composed of variables and composite variables. The terminology they have selected is meant to distinguish simple variables from more complicated sets of variables and relationships.

This is not to say that an uninstantiated value view has little or no value. The value view provides a framework that both guides efforts to instantiate it and organize knowledge. Properly done, the value view represents the state of our knowledge and can inform a variety of investment and policy decisions. For those wishing to explore parts of the landscape, the value view provides a point of departure and a checklist of what needs to be considered.

Because few experiments or data collection efforts are practical at the enterprise level, ultimately the value view is informed by observations or instrumentations of reality, the conduct of experiments, and mission-oriented models and analyses.

The next chapter discusses the nature of a process model or process view.

Informing the value view

CHAPTER 8

CONCEPTUAL MODEL: PROCESS VIEW

The process view, as introduced in Chapter 5 (Figure 9), organizes functions and processes, whether past, current, or future, into a small number of conceptual bins. The bin "situation information" represents the host of information-related assets and processes that sense, collect, process, protect, disseminate, and display information.[116] The product of these processes (information) provides data about the environment, including the effects of interest. This information is used as an input to all of the other process concepts. Figure 25 makes the key processes inherent in **C2** explicit.

The concepts depicted in Figure 25 are generic. They are focused on the functions that need to be accomplished, not on the way in which these functions are or could be accomplished. The boxes represent processes and activities that accomplish the functions involved. The arrows represent the relationships between the products of these processes and

[116] The word *information* is used here to include what has been defined as data, information, knowledge, understandings, and even wisdom. These terms are in fact *fitness for use*-based or relative to a situation rather than absolute.

other processes. For example, the function of *command* produces, among other things, the product of intent. *Command intent* is used as an input to *sensemaking* as it is moderated by *control* or more specifically by the processes that have been developed to accomplish the functions associated with *control*.

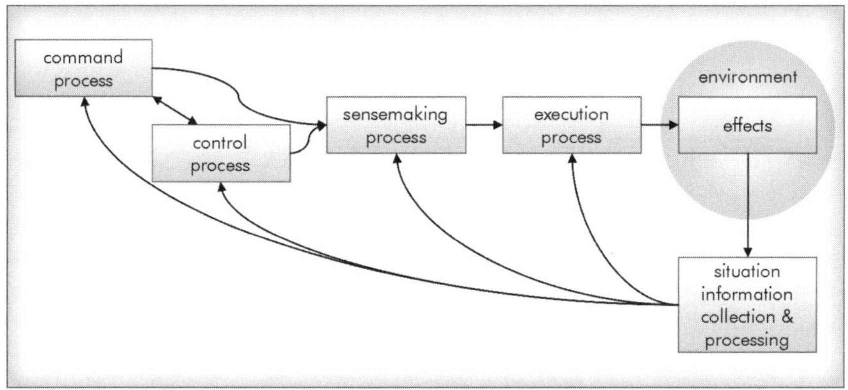

FIGURE 25. C2 CONCEPTUAL MODEL: PROCESS VIEW

A process is simply a series of actions or operations progressing to an end. A process view defines the entities, systems, resources, and the interactions among those actions or operations. Processes have inputs and outputs. A process view specifies the sequence of steps that transform the inputs into outputs and show how the outputs of one process feed into another. A process view implicitly or explicitly incorporates *time* or *sequence*—what happens in what order. There are metrics associated with each of the processes and with the relationships between and among them. In a process view, it is the actual products, not assessments of the products (their quality), that are the links in a process or event chain. This generic process view serves as an overall architecture for specific process instantiations. A specific process instantiation is one that represents a particular way of accomplishing a function and the specific product(s) produced.

Process instantiations

Researchers and analysts working on different aspects of **Command and Control** will develop specific instantiations of the processes of interest and importance to them. This section looks at three such instantiations. All of them are consistent with the basic **C2** Conceptual Model introduced in Chapter 5 and the Process View shown in Figure 25.

- The HEAT (Headquarters Effectiveness Assessment Tool) Model of Industrial Age **Command and Control**;
- The Network Centric Operations Conceptual Framework developed by the U.S. Office of Force Transformation and the OSD/NII Command and Control Research Program to examine **C2** as it is understood from that perspective, which dates from 2002; and
- A new process view developed for this book that builds upon the NATO SAS-050 product.

Each of these process models is actually composed of hundreds of individual variables and the metrics associated with them. However, the processes they include can be shown in somewhat less than full detail by focusing on an intermediate level, which is the approach adopted in this chapter.

HEAT MODEL[117]

The HEAT approach was developed in the early 1990s to examine the then-prevalent Industrial Age **Command and Control** processes recognized several components of the process. It is depicted graphically in Figure 26.

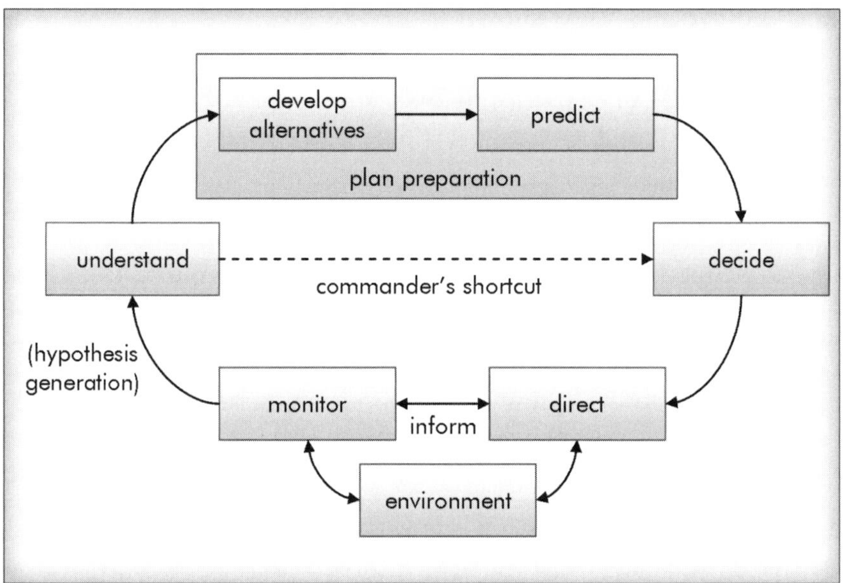

FIGURE 26. THE HEAT MODEL

- **C2** (seen as one function) always operates and can only be understood in the context of an operational environment, which is comprised of the physical environment (terrain, weather, man-made structures, etc.), a friendly

[117] For more on the Headquarters Effectiveness Assessment Tool, see:
Hayes, Richard E., Mark Hainline, Conrad Strack, and Daniel Bucioni. "Theater Headquarters Effectiveness: Its Measurement and Relationship to Size Structure, Functions, and Linkage." McLean, VA: Defense Systems, Inc. 1983.
Hayes, Richard E., Conrad Strack, and Daniel Bucioni. "Headquarters Effectiveness Program Summary." McLean, VA: Defense Systems, Inc. 1983.

and enemy situation, and a political, social, and economic context. For any given headquarters or set of headquarters, the forces reporting to them and cooperating with them, as well as adversary and neutral forces, are all part of the environment. The missions of those forces at any given point in time also form part of that environment.

- The purpose of a **C2** system is to bring or keep selected aspects of that operational environment within some desired boundaries. Those boundaries will vary by the type of conflict and the goals of the military forces. The relevant measures of force effectiveness are keyed to the factors being bounded. They may involve casualty ratios, terrain captured, destruction of an enemy's will or capability to continue the fight, or other specific goals. Hence, the **C2** system is understood in HEAT to be an adaptive control system.
- In order to perform its role successfully, a **C2** system must first *monitor* its environment, which is done through a variety of technical and human means, including reports from its own forces about their status and activity as well as intelligence, weather, and pre-established knowledge such as the order of battle and geography.
- The data and information arriving about the operating environment are combined into some *understanding* of the most salient features of the environment. These understandings form the perceptual basis for decisionmaking. They represent what those in the headquarters know (or believe to be true) and how they see the situation evolving over time.
- Given an understanding of the operating environment, the **Command and Control** system will generate *alternative courses of action* intended to achieve or maintain

control over those aspects of the environment considered most important for mission accomplishment.

- Each alternative course of action (which includes doing nothing or continuing to follow the existing plan) is assessed (sometimes formally, sometimes simply within the heads of commanders or other key personnel) and a mental *prediction* is made about its feasibility and the likelihood that it will generate a desirable future state.
- Based on this set of predictions (which can be formal or intuitive), the commander makes a *decision,*[118] a choice between the alternative courses of action.
- Decisions must be translated into *plans* or *directives*[119] that both inform the elements of the force of the goals of the effort, their role in achieving those goals, the resources or assets available to them, boundaries and other control measures to de-conflict or integrate the efforts of the force elements, the schedules to be followed, and recognizable contingencies under which goals, assets, boundaries, or schedules will be changed.
- These directives must then be delivered[120] to the elements of the force, which act to implement the plans and influence the operating environment.
- This, in turn, along with enemy actions and other changes in the operating environment (e.g., weather, movement of refugees) keep that environment dynamic, which means that the **C2** system must restart the cycle in order to monitor these new developments.

---

[118] HEAT assumed that a headquarters was commanded by a single commander and that this commander made the selection among alternatives that were generated and preliminarily assessed by the headquarters staff. The model presented in this book does not make this assumption.

[119] HEAT assumed the nature of these plans was to follow the then-best practice.

[120] HEAT assumed "push" within the chain of command as the mechanism for information dissemination.

- This cycle may be repeated as the situation changes or it may be "short circuited" when the operating environment changes in a way that was predicted so that contingency plans can be triggered and implemented promptly.

The HEAT approach, like its more tactical cousin the OODA (Observe, Orient, Decide, and Act) loop, proved useful and has been successfully applied to Joint, Army, Navy, Air Force, and coalition operations. However, it has proven too thin to support analyses of Information Age operations. Moreover, it tends to reflect a cyclic **C2** process in which forces are always acting on guidance that is somewhat out-of-date and higher headquarters are always relying on information that is somewhat out-of-date because of the time required to move information and guidance up and down through the hierarchical **Command and Control** system. This is illustrated in Figure 15 (Chapter 6), with the shaded areas indicating the time-lagged movement of both information and guidance.

The Network Centric Operations Conceptual Framework (NCO CF)

The NCO Conceptual Framework was developed as a mechanism for understanding the implications of the NCO tenets.[121] It is a tool for education about NCO and a basic structure designed to help researchers organize their work and apply comparable metrics across projects and domains. Figure 27 provides a top level view of the NCO CF.

[121] Network Centric Warfare Report to Congress. 2001.

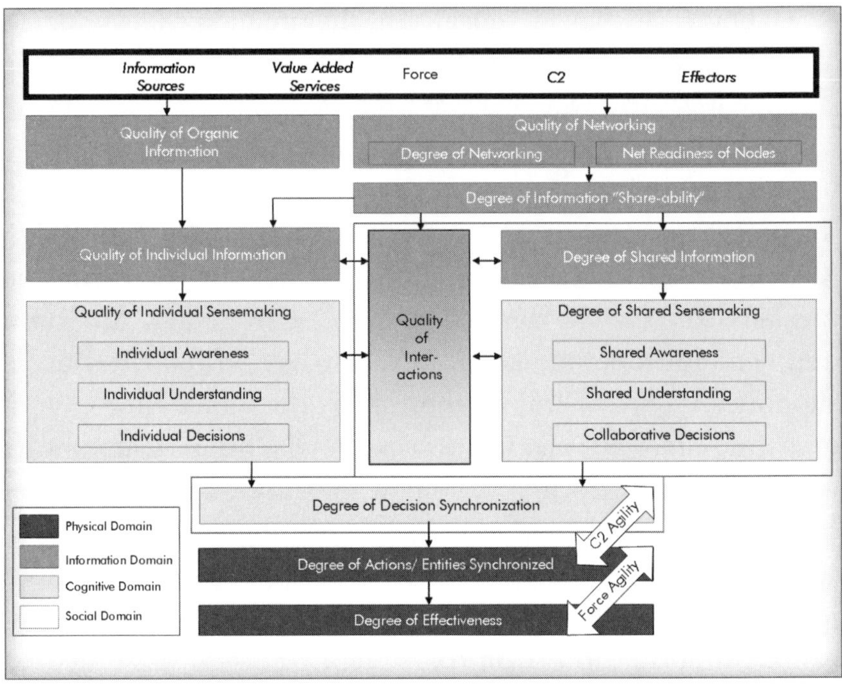

FIGURE 27. NETWORK CENTRIC OPERATIONS CONCEPTUAL FRAMEWORK (TOP LEVEL)

While the NCO CF assumes an operating environment and explicitly discusses one in its text,[122] the topic is not explicitly addressed in the graphic used to convey the components and relationships of the **C2** process. Rather, that graphic initiates the process based on the force, which is composed of four elements: (1) information sources (which can be sensors, reports, open source news such as CNN, human intelligence, or any other originator of data or information), (2) value-added services (such as processed intelligence or integrated reports formatted for ease of understanding), (3) the **Command and Control** function (which is understood to be elaborated in the

[122] Office of Force Transformation, "Network Centric Operations Conceptual Framework, Version 1." 2003.

rest of the graphic), and (4) effectors (anything that influences the operating environment such as kinetic weapons or psychological warfare broadcasts).

The NCO CF process model highlights the fact that **C2** must be understood to involve four different domains: the physical (radios, satellites, wires, and other hardware and the software required for their operation), information (the substantive contents of the **C2** system such as the location, status, and activities of the relevant entities to the extent that they are known in the information systems), the cognitive (what is in people's heads), and the social (relationships between and among individuals and groups of people) domains. The recognition of these four interrelated domains is crucial to understanding the requirements for successful Network Centric Operations. This makes it clear that the **C2** processes are not purely mechanistic, but in fact are highly dependent on human capabilities and behaviors as well as the nature of the interactions among individuals, teams, and organizations.

The NCO CF's information domain processes highlight the differences between data, information, knowledge that individuals and organizations possess or have access to based on their own senses, the sources they control (organic assets), and the information they receive as a result of being connected to the network. Examining the "quality of the network"[123] allows processes that involve more or fewer players and more or less interaction and information gathering capability to be distinguished. The ideas of a richly networked information infostructure (lots of capacity, lots of information, lots of appli-

---

[123] *Network* is used here to refer to the distributed information environment or infostructure, a collection or system of systems, and not to "network" in its broad social domain context.

cations that make it easy for users to benefit from that capacity, high levels of information assurance) and "net-ready users" who are both plugged into the network and prepared (cognitive capacity, training, etc.) to use it efficiently are built into this part of the process model.

Recognizing the distinctions between the cognitive and the social domains allows the process version of the NCO CF to stress the difference between what individuals in the force (commanders, key staff, etc.) know and do from what the crucial groups of key personnel (commanders across levels, leaders across functional areas, commanders and their staffs, communities of interest) know and do within the C2 context. Individuals are seen as having access to different levels of information, which gives them personal awareness and individual understandings (understandings involve combinations of what they are aware of and their existing mental models and knowledge so they understand cause and effect as well as temporal dynamics and thus a sense of what futures are possible and how they might be influenced). Decisions are considered to be choices among alternatives. The nature of the decisions required is determined in large part by the perceived distribution of responsibility and authority. Taken together, these three cognitive processes (awareness, understanding, and decisionmaking) constitute "sensemaking."

The NCO CF also recognizes that individuals seldom work in isolation when carrying out Network Centric Operations in the Information Age. Hence, the process model recognizes the quantity and quality of interactions between individuals as important, if not critical, for NCO. As a consequence, the NCO CF takes into account the opportunities available for sharing information and for collaboration that exist in a particular process instantiation of NCO and also considers the extent to which these activities actually occur. Thus it makes a clear distinction between potential and its realization.

At the same time, the NCO CF traces a parallel track for the social and cognitive activities of the teams and groups involved in network-centric C2. Here again, the trace runs from the degree to which shared information is available for sensemaking including awareness, understanding, and decisionmaking. However, this process model also notes that these "shared" cognitive activities (strictly speaking, of course there are no shared cognitions since there are no shared brains) can result from collective knowledge (what anyone in a group or on a staff knows) to partially shared knowledge (what two or more members know in common) to the union (what everyone in the group or team knows). Moreover, these distinctions can be used to represent different work processes and may imply different outcomes or requirements for time and effort in order to reach shared awareness, understandings, or decisions.

Whether the products of individual or shared sensemaking, decisions within the force are seen in the NCO CF as characterized by some degree of synchronization (purposeful arrangement in time and space[124]). The level of synchronization is recognized as a metric worth capturing. This part of the process model is important at least in part because it is causally linked to the degree to which the actions of different elements of the force are themselves synchronized or the actions of any given element are synchronized over time. In turn, synchronization is seen as influencing the degree of the effectiveness of the force.

---

[124] *Space* is used here to mean more than physical 3-dimensional space or geo-reference. Rather, it refers to the four domains.

Finally, the NCO CF incorporates the concept of agility, as defined in *Understanding Information Age Warfare,*[125] and highlights its importance in the analysis and assessment of Network Centric Operations. *Agility,* as used here, is made up of six components: robustness, resilience, responsiveness, flexibility, innovation, and adaptation.[126] These are seen as important not only in the **C2** arena, but also in the performance of the force itself in the operating environment.

The Network Centric Operations Conceptual Framework was developed over a period of more than 3 years. It has proven effective as a way to analyze case studies of forces with some network-centric attributes, to teach NCO to analysts, and to organize research efforts focused on experimentation and concept development. It has been applied to situations in Iraq and Afghanistan, as well as peacekeeping and humanitarian operations. However, like HEAT, it represents only one class of **C2** processes, and not the universe of possible approaches.

NATO Conceptual Model for Understanding C2

While HEAT and the NCO CF were developed to capture the specific processes in a particular **C2** Approach, the conceptual model (CM) developed by the NATO SAS-050 Working Group was deliberately designed to support research, analyses, and insights into all (or at least the broadest possible variety of) alternative approaches to the command function and the control function. This CM explicitly recognizes the difference between the value view, the process view, and the reference model (or event view) in the range of functions involved. That

[125] Alberts et al., *Understanding Information Age Warfare.* p. 197.
[126] Alberts and Hayes, *Power to the Edge.* Chapter 8.

model remains immature at this time and will require further work.[127] However, it has reached the stage where it can be used as a reference model. At this writing, it includes more than 300 specific variables and posits more than 3,000 relationships among them. SAS-050 has provided major insights into the process model needed to understand the universe of possible **C2** Approaches. While no consensus exists as to the best way to represent a top level view of the SAS-050 reference model, the graphics and discussions that follow have been heavily influenced by SAS-050.

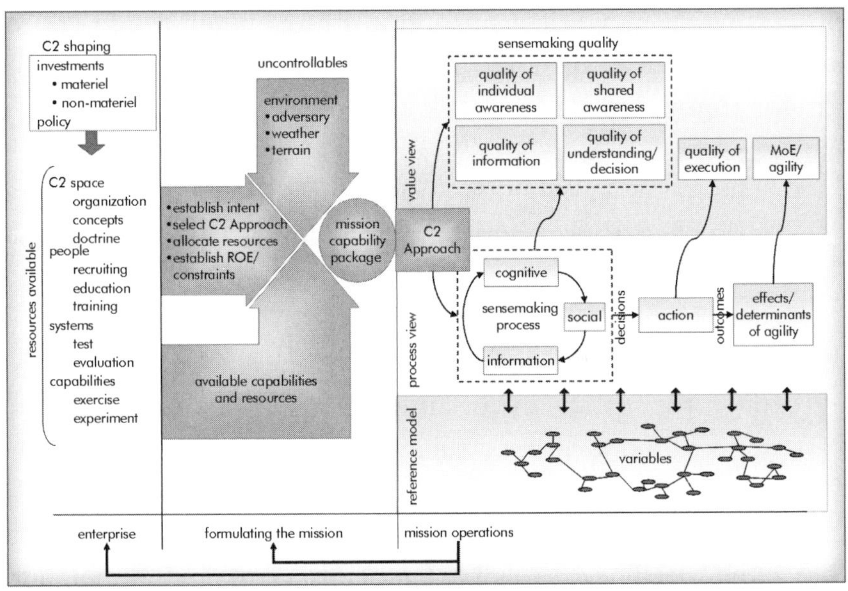

FIGURE 28. NATO CONCEPTUAL MODEL FOR UNDERSTANDING C2

This approach differentiates the **C2** aspects of the enterprise in which the force itself is created (in the U.S., by the Department of Defense) and the processes by which a mission and the force

127 NATO SAS-050 Final Report. 2006. Available at: <www.dodccrp.org> March 2006.

assigned to it are shaped into a mission capability package from the actual "mission operations" as the force is employed.

The enterprise is not represented in detail, but is recognized to be a combination of policies and investments that create and distribute the assets available in order to create capabilities. While these efforts do receive inputs over the longer term (years or more) from operations in the field, they do not contain the **C2** processes of greatest interest in this effort. Similarly, the conceptual model recognizes that nations (or coalitions and alliances like NATO) go through an important process when they select missions and formulate the forces to accomplish those missions. Three major elements are seen as driving that process:

- The forces (capabilities and resources) available, particularly when those contributing the forces have other obligations at home, within their regions, and around the world;
- The uncontrollable factors that shape the mission and the force such as the weather, terrain, operating distances, and facilities available, as well as the perceived capabilities and intentions of the adversary and other relevant actors; and
- The specific controllable factors that can be manipulated by those organizing the effort, such as the intent associated with the mission, the specific resources allocated to the effort, and the constraints recognized, including the need to honor the neutrality of some parties, and the rules of engagement.

In this process model, one of the most important decisions to make in order to represent an existing or a possible mission capability package is the selection of a **C2** Approach. As illustrated in Figure 11 (page 75), a **C2** Approach[128] consists of (1) the way decision rights are allocated, (2) the patterns of interaction that are enabled, and (3) the distribution of information across the elements of the force. While the **C2** Approach may change over time (for example, as the force moves from crisis management to combat to stabilization operations) or differ across function (logistics and artillery may employ very different approaches), choices made on these three fundamental dimensions are profound decisions with far-reaching implications for the overall **C2** process.

As shown in Figure 28, taken together, the factors that drive the formulation of a mission ultimately result in the creation of a mission capability package that includes all of the capacities (and limitations) of the elements of the force. This MCP represents the forces involved in the mission at any one point in time. However, it may also change based on either changes in the mission or adaptation to better carry out that mission.

In the NATO model, most of the focus is on the "mission operations" phase, which clearly includes both a value view (where the value chain of the model resides) and the underlying reference model where the hundreds of relevant variables reside and are given values if and when the CM is organized into a simulation or operating model.

Because **Command** and **Control** are each processes in and of themselves, the conceptual model for understanding **C2** must

[128] The concept of a **C2** Approach space adopted by SAS-050 was created, in larger part, for early drafts of this book.

be understood to be a "meta-process" composed of three more specific sub-processes, including:

- Information processing, which accepts information from and about the operating environment and converts it into useful data or products;
- The sensemaking process by which information is converted into awarenesses, understandings, and decisions at either the individual or some shared level; and
- An implementation process by which decisions are converted into actions.

Each of these, in turn, is made up of other sub-processes. For example, information processing includes at least tasking, posting, search, discovery, fusion, retrieval, integration, information assurance, and collaboration. Sensemaking includes cognitive processes, social processes, and some, such as decisionmaking, that are both cognitive and social. Implementation includes planning processes as well as the actual movement of objects. Moreover, these processes interact—as new information becomes available it impacts both sensemaking and implementation: sensemaking generates new taskings for information processing and implementation provides feedback on plans and generates new information flows.

However, because a process view consists of a specific sequence of tasks and temporal information, this view cannot be constructed in detail for the generic model. These details can only be deduced after the **C2** Approach or set of relevant **C2** Approaches have been determined. While the HEAT model of an Industrial Age **C2** Approach and the NCO CF model of a specific Information Age **C2** Approach permit specification of the processes that they embody, the more generic CM offered here will not support such analyses. In its place we have provided the following chapter on "Influences" that discusses key causal relationships inherent in any **C2** Approach.

CHAPTER 9

INFLUENCES

INTRODUCTION

While the NATO Conceptual Model for understanding **C2** is too generic to support the specification of a particular instantiated process model, it is rich enough to create and analyze networks of influence. The term *influence* is used here in its most basic meaning: the power to affect a person or process. Note also that what follows are illustrative networks of influence, not formal influence models, which would require specification of the valence, strength, and conditional relationships between variables. These networks of influence could be used as the basis for the experimentation and observation necessary to construct such influence models and capture them in formal tools such as Systems Dynamics, but the state of the art at this writing does not provide enough knowledge for that purpose. Moreover, that type of modeling would be easier if a particular **C2** Approach were being instantiated.

Significant Influences Involving Collaboration

Collaboration plays a central role in almost all **C2** Approaches. Some systems, such as an Industrial Age hierarchical system, stress individual responsibility and accountability and focus their efforts on supporting and empowering those individuals (primarily commanders at all levels). Other (Information Age) approaches, such as those described as "network-centric" or "network-enabled," stress the role of collaboration in improving performance. Indeed, one of the key issues in selecting a **C2** Approach is defining the appropriate use and limits for collaboration.

The significant influences involving collaboration are depicted graphically in Figure 29, which tends to flow from the top to the bottom. Three aspects of collaboration are chosen as focal points: the nature of collaboration, the speed of collaboration, and the likelihood of successful collaboration. The first is a composite variable of the myriad factors surrounding the **C2** Approach. The second means simply how much time is consumed by a given collaboration interaction. The third deals with the extent to which the goal(s) of any given collaboration will be achieved. As discussed below, collaboration (or the lack of it) plays several roles in any **Command and Control** Approach, often more than one simultaneously. For example, collaboration designed to improve the information available almost always results in changes in awareness. Hence, "successful" collaboration is typically multi-dimensional. The three focus variables are shaded darkly in Figure 29.

Significant influences involving collaboration

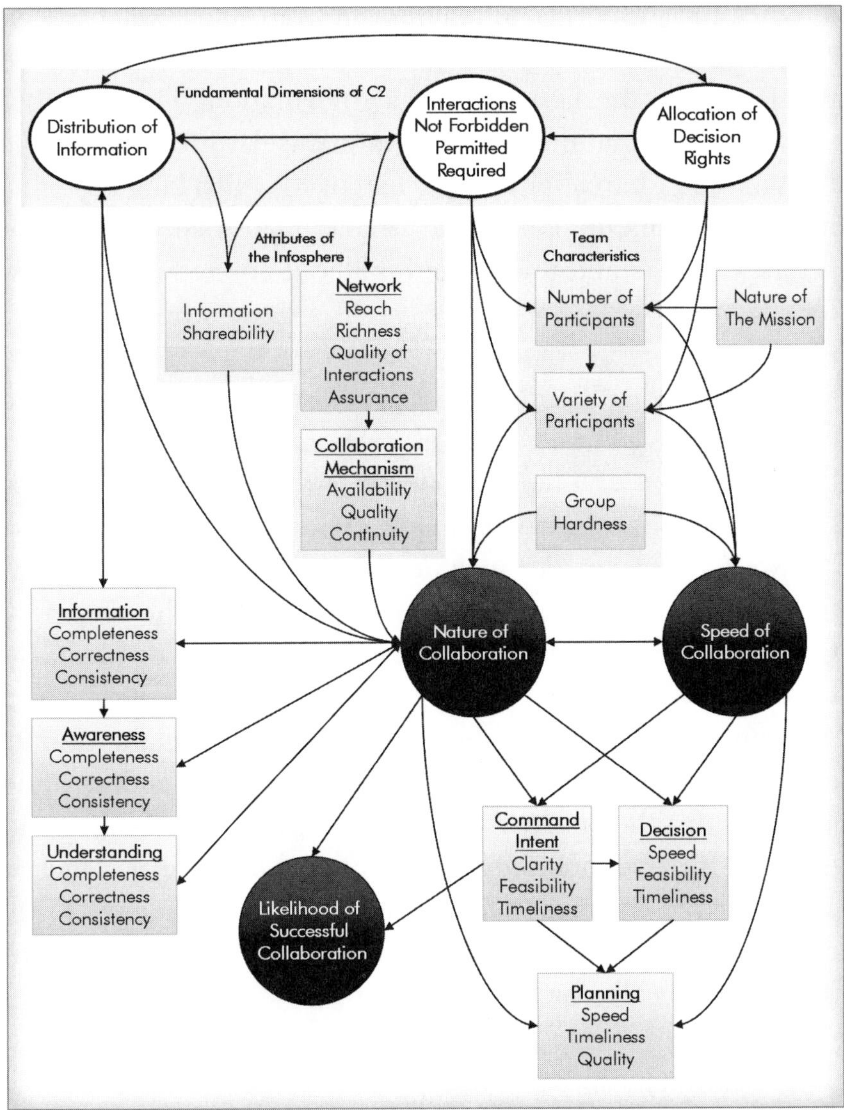

FIGURE 29. COLLABORATION: SIGNIFICANT INFLUENCES

The beginning point for the network of analytic influences is the set of three fundamental factors that determine the character of all **C2** Approaches: the Allocation of Decision Rights, Interactions Permitted, and the Distribution of Information. While the dimensions of a **C2** Approach all interact with one

Significant influences involving collaboration

another, the strongest causal links run from Allocation of Decision Rights to the set of Interactions Permitted and from those two factors to the Distribution of Information. The Distribution of Information is also influenced directly by the Information Shareability or the extent to which language, formats, and semantic consistency make it possible for different actors to access and understand material entered into the system by others.

The Interactions Permitted also has a strong influence on the characteristics of the technical network available to support the **C2** processes. The characteristics of the distributed infostructure, labeled "Network," are contained in a box. These Network characteristics include Reach (breadth of participation), Richness (the quality of the contents of the network), Quality of Interactions (the breadth and depth of media in which people and systems can interact on the network), and Assurance (the extent to which the network is available and its contents are secure). These Network characteristics determine (along with other factors not shown such as the quality and training of personnel) the Collaboration Mechanism, which is characterized in terms of its Availability, the Quality of the collaboration enabled (data, voice, images, etc.), and the Continuity of the collaboration it can support.

The other major influences on the success of collaboration are the Number and Variety of Participants, Group Hardness, and the Nature of the Mission. There is a simple relationship between the Number of Participants and the Speed of Collaboration. All other things being equal, more participants require more time for collaboration. The Number of Participants is also one influence on the Variety of Participants, which means that with a greater diversity of individuals involved, more numerous and diverse perspectives (people with different experience, training, roles, etc.) will be involved. The Variety of Participants both slows the Speed of Collaboration (they need time to communicate across organizational, cultural, and other boundaries) and also affects the Nature of Collaboration, which in turn impacts the Likelihood of Success. Except on very simple problems, involving more participants improves the likelihood of success, both by reducing the likelihood of "groupthink" and also by increasing the chances that relevant experience and training are available to contribute to success.[129]

However, at least two other factors will also impact collaboration. First, Group Hardness (the extent to which the individuals involved in the collaboration have worked together in the past on similar problems and developed trust and a common language and methods for dealing with issues) both increases Speed of Collaboration and affects the Nature of Collaboration. Indeed, hardness is the major tool for overcoming the loss of speed that occurs when more participants become involved and different perspectives need to be integrated. Hardness also enables the group to improve its performance. This is one major reason that military forces value unit training and field exercises: they improve people's ability to work together. Similarly, when organizations (joint, coalition, interagency, international, public, and private) work together over time, they become hardened and improve both their efficiency and effectiveness. The other key factor that directly impacts the Nature of Collaboration is the Distribution of Information across the people and nodes. All other things being equal, the richer that distribution is (the greater

[129] See: Druzhinin and Kontorov, *Concept, Algorithm, Decision.*

the ability to capture and share information), the more likely it becomes that collaboration will be successful and (indirectly) be accomplished more quickly. Of course, the limits of human cognition apply, so very rich information distribution may also require filters, frames, or other tools that manage the load on individuals. Speed of Collaboration is also impacted by, and impacts, the Nature of Collaboration.

The profound role of collaboration in **C2** becomes obvious when its multiple roles are understood. As indicated by the boxes labeled Information, Awareness, and Understanding as well as the two-way arrows that connect them to the Nature of Collaboration, the process of collaboration has a rich and resonant impact on key activities. Moreover, these interactions are richly interrelated and simultaneous. They also occur both in dealing with individuals and groups, so the three boxes include not only what is available to any one person, but also Shared Information, Shared Awareness, and Shared Understanding.

There is a natural flow from Information (what is known within the system) to Awareness (what individuals and groups understand the situation to be and be becoming at any point in time) to Understanding (Awareness plus cause and effect relationships and temporal dynamics such that individuals and groups foresee alternative futures or patterns of futures). Successful collaboration is the means by which *Individual* Information, Awareness, and Understanding are converted into *Shared* Information, Awareness, and Understanding. In fact, these factors are closely coupled and virtually simultaneous.

As illustrated in the figure, the specific variables most strongly influenced by the Nature of Collaboration are Completeness, Correctness, and Consistency, though other factors such as

Currency are also likely to be impacted. Here, again, these influences chain across Individual and Group as well as Information, Awareness, and Understanding. Note also that these influences are mutual; better collaboration improves and is improved by better Information, Awareness, and Understanding at both the individual and group level.

The most significant influences of Speed of Collaboration and the Nature of Collaboration in later stages of **C2** are their impact on Command Intent, Decisions, and Planning. Successful collaboration will impact the Clarity of Command Intent because more of the actors will have been involved in developing Command Intent, both giving them some prior knowledge of it and the rationale underlying it and also helping to ensure that semantic interoperability is higher. In addition, the Nature of Collaboration will also impact whether the Command Intent is Feasible because it is very likely to have been developed using more perspectives and expertise. The Timeliness of Command Intent will also be influenced by the Speed of Collaboration.

The Nature of Collaboration also improves Decision Feasibility. More successful collaboration implies that decisions, like statements of intent, will take into account a broader range of experience, expertise, and perspectives. Precisely the same logics indicate that Planning (whether it is deliberate, hasty, or on the fly) Quality, Speed, and Timeliness will be influenced by the Speed of Collaboration and the Nature of Collaboration.

In summary, the role of collaboration (working together for a common purpose) in **C2** is not always understood or appreciated. This is one of the cornerstones of any **C2** Approach and the extent and role of collaboration is one of the factors that

Significant influences involving collaboration

most strongly differentiates Industrial Age from Information Age approaches. Research into how to best organize and use collaboration is one of the areas where future research (observation and instrumentation of war games, exercises, and operations) and experimentation should focus. A great deal is known already from the small group, business, and group dynamics literatures, but little work has been done on military organizations, interagency operations, coalition operations, or public-private cooperation in humanitarian and reconstruction operations.

Significant influences involving agility

As reflected in Figure 30, characterizing the factors that influence agility requires a different mindset than that required to identify the factors that influence collaboration. The major reason that an alternative approach is needed is the extremely close coupling between and among the factors that influence agility. Four sets of factors are involved:

- The characteristics of the mission capability package under discussion;
- C2 qualities;
- The components of agility; and
- The mission challenges that drive the need for each of the elements of agility.

Virtually (if not literally) every possible influence within and across each layer is relevant. Hence, diagramming the individual influences would yield a virtually incomprehensible image.

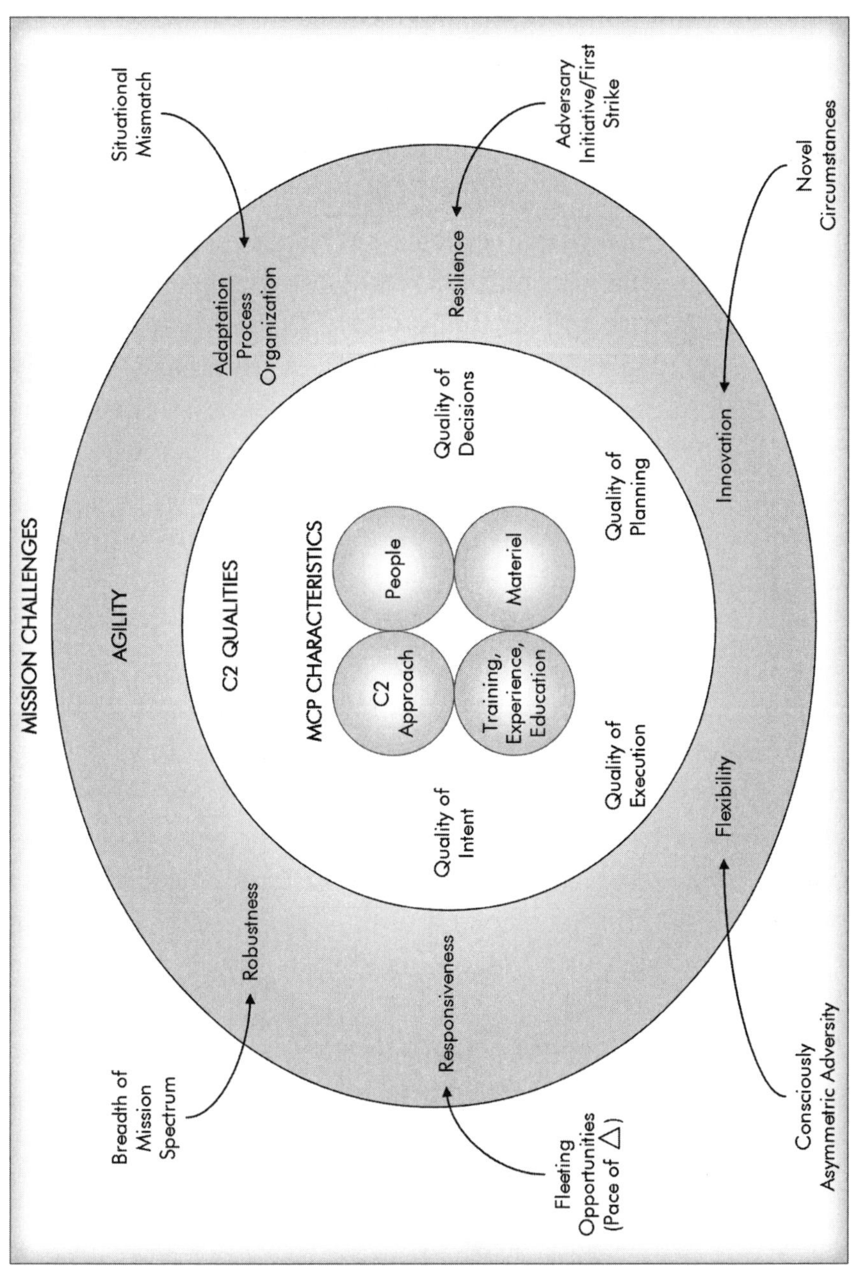

FIGURE 30. AGILITY: SIGNIFICANT INFLUENCES

Significant influences involving agility

The core level of the graphic involves the fundamental characteristics of the mission capability package (or packages). This means virtually every aspect of the force: the *materiel* (weapons, infostructure, logistics for deployment and sustainment, and so forth), the *people* within the force, their level of *training, experience, and education*, and the **C2** *Approach* (or set of approaches) adopted. These factors interact with one another to determine the characteristics and attributes of MCPs and their respective capabilities. Figure 30 organizes these sets of factors into layers. The innermost layer or core corresponds to two of the NATO *Code of Best Practice for C2 Assessment* classes of measures of merit: Dimensional Parameters (characteristics of people and systems) and Measures of Performance (the way the systems perform their functions).

This inner core strongly influences the qualities of several key measures of **C2** quality. The most relevant of them are shown in the diagram: quality of command intent, quality of decisions, quality of planning, and quality of execution. These four measures are closely interrelated; strength or weakness in any one of them will have an immediate and measurable impact on the others. This layer corresponds with the NATO *COBP for C2 Assessment* Measures of **C2** Effectiveness.

Agility has developed into one of the most important concepts in assessing alternative **C2** Approaches. While this term is used loosely in most of the literature, we have identified its six key elements in *Power to the Edge.*[130] As a multi-dimensional concept that interacts with the operating environment and the internal workings of any specific instantiation of **C2**, agility is actually composed of:

---

[130] Alberts and Hayes, *Power to the Edge*. pp. 127-128.

Significant influences involving agility

- Robustness – effectiveness across a range of tasks, situations and conditions;
- Resilience – the ability to rebound from damage or misfortune;
- Responsiveness – the ability to act within windows of opportunity;
- Innovation – the ability to do new things or old things in new ways;
- Flexibility – the ability to accomplish missions in multiple ways; and
- Adaptation – the ability to alter process and organization to improve effectiveness or efficiency.

Agility presumes effective actions and implies a degree of self-synchronization. This further implies that the elements of the force behave reliably and predictably. C2 quality provides the basis for agility. The relative weaknesses and strengths associated with C2 strongly influence the level of agility possible in each of the six key areas. At the same time, the elements of agility also influence one another. For example, greater flexibility requires greater innovation. Similarly, greater robustness depends in no small measure on resilience, responsiveness, flexibility, and adaptation. In fact, when a detailed review of these influences was undertaken, all of these elements were rapidly shown to be related to one another. Hence, while it remains possible to specify definitions and metrics for the elements of agility, the concept must be considered holistically. In that sense, it remains a key component of networked or complex adaptive systems.

Outside the agility layer in Figure 30 lies Mission Challenges, which represents those aspects of the operating environment that create the need or requirement for each of the elements of agility.

- *Robustness* is a necessary response to the need to operate across the mission spectrum. Optimizing against any one adversary or class of adversary would mean a lack of preparation against others known to be serious threats to U.S. national security interests.
- *Responsiveness* is required to deal with fleeting opportunities, whether in the tactical arena (fleeting targets) or at the operational and strategic levels (windows of opportunity in dynamic situations). As the pace of change has increased across the globe and adversaries have become more adept, the need for responsiveness has grown.
- *Flexibility* is increasingly required because adversaries (particularly terrorists and insurgents) are consciously studying our doctrine, practices, and experiences in order to improve their chances of success. This means that we will need multiple ways to succeed so that their ability to thwart some of our approaches will not prevent mission accomplishment.
- *Innovation* is the natural response to efforts by our adversaries to place us in unfamiliar situations or exploit our predictability. By doing old things in new ways or entirely new things, we reduce our forces' vulnerability.
- *Resilience* will be essential because we cannot assume that adversaries will not be able to strike first or otherwise seize the initiative. Hence, we will suffer casualties and have our operations disrupted. This cannot be allowed to lead to failure.
- *Adaptation* (changing our processes and organization) is needed when we find ourselves in a "situational mismatch" such as the battle against the Taliban in Afghanistan. In such cases, our existing organizations and processes must be adapted in order to allow effective mission execution.

Significant influences involving agility

Like collaboration, agility has not been well-understood within the **C2** community. As discussed here, a better understanding of these two concepts and what influences them is crucial to understanding alternative **C2** Approaches and their implications.

In this book, we have, in essence, stipulated the need to go back to basics, to put aside what we think we know about what **C2** is, because for many this amounts to how **C2** has and is currently accomplished. It is often said that it is far easier to learn something new than to forget something old. The concluding chapter, The Way Ahead, looks at the journey before us and highlights the major tasks and challenges that we face.

CHAPTER 10

THE WAY AHEAD

OVERVIEW

Developing a better understanding of **Command and Control**, particularly of the class of **C2** arrangements and processes that work well with network-centric and coalition operations, is on the critical path to DoD transformation. To improve our current understanding, we must not only improve our models, improve our ability to measure key variables, and improve our ability to conduct analyses and experiments, but we also need to push simultaneously both the state of the art and the state of the practice of **Command and Control** itself. The state of the practice is generally understood to be on the critical path, but the connection between understanding **C2** (the state of the art) and improving the practice of **C2** is not as widely recognized. This concluding chapter will explain this connection and the need for a multi-pronged effort, one that simultaneously involves both the state of the art and the state of the practice.

Additionally, we must change the approach and metrics we use to evaluate investments as needed. The reason for this needed change and the nature of the new approaches and metrics required will be discussed later in this chapter.

Synergies: Art ↔ Practice

Efforts that simultaneously address the state of the art and the state of the practice of **Command and Control**, and invest in improvements to the state of the practice of analysis, experimentation, and measurement, can generate significant synergies.

Experimentation is an important source of ideas that contribute to the state of the art of **Command and Control**. Experiments also provide empirical data that contributes to improving our analyses and our understanding of **C2**. Improving the state of the art of **C2** serves as a source of possibilities for practitioners to improve the state of the practice by applying theory to practice. Advances in the state of the art need to be generalized and incorporated into our understanding of **C2**. Hence, the state of the art provides "requirements" that need to be understood.

At the same time, developments in the field that improve the state of the practice provide opportunities to collect valuable data and to test instrumentation and measurement approaches. Innovations in the field not only improve the state of the practice but contribute to improving our understanding of **C2** by generating a set of possibilities that need to be systematically explored. Improvements in understanding **C2** provide knowledge to practitioners, ideas that can improve the state of the art, focus for experiments and analyses, and requirements in the form of the variables and relationships that need to be measured and explored. Improvements in the state of the practice of analysis improve the analyses we undertake. These analyses provide knowledge that enables us to improve our understanding of **Command and Control**. Finally, improving the state of the practice of measurements improves our ability to measure key variables and relationships, helping experimenters by giving them the tools they need and analysts by providing better data. Figure 31 depicts these synergies.

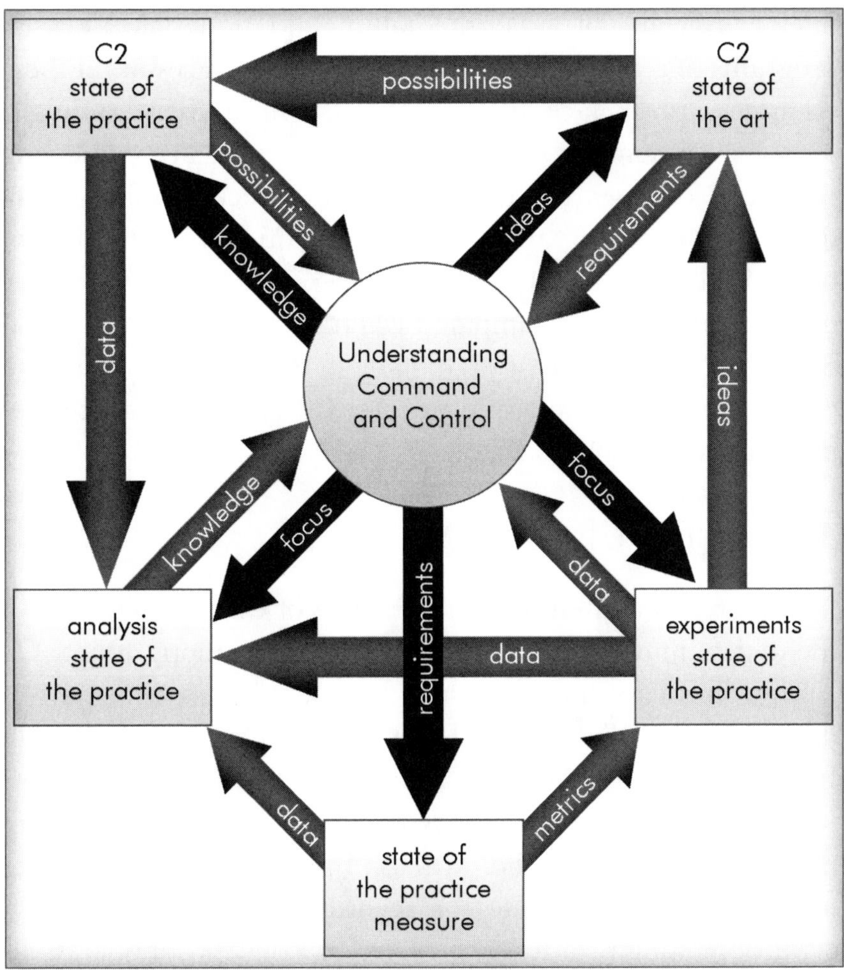

FIGURE 31. SYNERGIES: ART ↔ PRACTICE

A failure to invest and harvest the advances made in any of these areas diminishes our ability to undertake these activities, makes them less productive, and retards progress. Thus, the way ahead involves investments in each of the areas depicted in Figure 31 and the creation of a mechanism for sharing possibilities, data, ideas, knowledge, metrics, measurement techniques and tools, and most significantly, understanding.

Establishing a partnership between those who design and conduct campaigns of experimentation and undertake focused programs of research and **C2** practitioners is necessary if these synergies are, in fact, to be realized. Thus, when crafting a plan for campaigns of experimentation and accompanying programs of research, there must be a section of the plan devoted to engaging and involving practitioners at the beginning of the process rather than waiting, as is so often the case, until the phase of the campaign when demonstrations are conducted. Practitioners need to be offered an opportunity to contribute to the exploration of the **C2** Approach space, not just shown the "solution."

The Critical Path

In recent years, interest in the practice of **C2**, development of theoretical understandings of **C2**, and explorations of new **C2** Approaches have increased significantly. This, in large part, is due to an appreciation that DoD transformation is, in fact, an Information Age transformation and that the need for new ways of doing business means that new **C2** Approaches are required. New approaches are also required to adapt existing mindsets and practices to the security challenges of the 21st century, as well as to support emerging concepts of operations (e.g., network-centric and effects-based). As a result, individu-

The critical path

als and organizations around the world are engaged in a wide variety of research and analysis activities. They are producing useful data and creating important bits of knowledge.

It is vitally important that these **C2** research and analysis activities are not only supported but also expanded. Equally, if not more important, is the urgent attention that needs to be paid to creating the conditions necessary to get the most out of the **C2** research and analyses that are undertaken. Only by making the most of the investments we make can we hope to accelerate progress.

Research, analysis, and experimentation efforts are more efficient and more effective if they can build upon data that has already been collected and knowledge that has already been created. Clearly, taking full advantage of available data, research and analysis findings, and the existing body of knowledge requires that individuals and organizations be aware of what is available. Progress depends then on the level of shared awareness in the **C2** community.

However, simply knowing that some data was collected or that some analysis or experiment was done, does not, in and of itself, make these data and findings useful. The utility of the data collected depends upon:

- the existence of metadata,
- the relevance of the metrics used,
- the appropriateness of the instruments, and
- the conditions under which the data were collected.

The value of analyses and experiments depends upon the quality of their formulation and the extent to which the effort adheres to the NATO *Code of Best Practice for C2 Assessment.* Similarly, the value of the experiments that have been done depends on both how well they were designed and conducted. Both of these are a function of the degree of adherence to the *Code of Best Practice for Experimentation.* The value of the analyses and experiments that are conducted also depends on the quality of the conceptual and working models employed. These, in turn, depend on the quality of an evolving **C2** conceptual reference model that is, in the final analysis, a community effort. Thus, the prevailing state of the practice of analysis and experimentation determines the rate of progress as much as the degree to which the community has shared awareness of what exists, what is ongoing, and what is planned.

Community priorities

Understanding **C2** requires a community effort. It requires increased collaboration and cooperation between and among individuals and organizations who are interested in defense transformation in general, and specifically it requires new **C2** Approaches that anchor coevolved network-centric mission capability packages. Three areas warrant immediate, priority, and sustained attention:

- improving the **C2** conceptual reference model,
- promulgating, adopting, and using the codes of best practice, and
- improving the quality and dissemination of data, findings, and instrumentation.

Improving the C2 conceptual reference model

There have been two major efforts to develop a conceptual model that can be used to organize existing knowledge, focus research and experimentation, and support analyses related to an Information Age transformation. The first, sponsored by OSD (a collaboration of the Office of Force Transformation and the Command and Control Research Program in the Office of the Assistant Secretary of Defense for Networks and Information Integration) used the tenets of Network Centric Warfare as a point of departure for constructing a conceptual framework that could be used to structure case studies and convey what transpired in a systematic manner that facilitates comparisons between traditional and network-centric approaches to operations. The second effort, under the sponsorship of the NATO Research and Technology Organisation's Studies, Analysis, and Simulation Panel (research group SAS-050), independently[131] developed a **C2** conceptual model designed to facilitate the exploration of new, network-centric **C2** Approaches while, at the same time, improving our ability to analyze traditional approaches. This initial version of the NATO **C2** reference model was then validated by (1) applying the NATO *COBP for C2 Assessment* model to a case study to assess its utility (ability to support problem formulation), (2) conducting an extensive literature search to identify variables that were found to be relevant to **C2** and its relationship to operations, and (3) comparing the NATO model to the

[131] These efforts were largely independent although key members of the NATO group including its chairman participated in both efforts. However, the chairman insisted that the NATO group start with a clean sheet of paper and build their model from the experience of the participating analysts who came from both NATO and non-NATO countries.

OSD conceptual framework. As a result, the current version of the NATO C2 reference model represents the best thinking of a set of international experts[132] and provides the community with a conceptual model to employ in research, analyses, and experiments, and a firm foundation to build upon.

The NATO C2 Conceptual Model should be used as a source of ideas and a checklist to help ensure that research, analyses, and experiments consider all of the variables and relationships that are relevant to their efforts. Furthermore, it is important that those who have suggestions (additional variables, additional or modified relationships, and related metrics) contribute their ideas. The CCRP is committed to maintaining this model for the benefit of the community.

Promulgating, Adopting, and Using the Codes of Best Practice

Many years of effort have been devoted to the development of the three codes of best practices currently available.[133] Their value is clearly a function of the extent to which individuals and organizations are committed to adopting and adhering to them. As a result of the increased use of these codes of best practice in a variety of analyses and experiments, there will be a rich set of ideas about how to improve these codes. The CCRP Web site will host a forum for discussions focused on sharing information about the application of these codes to specific analyses/experiments and collect ideas for improve-

[132] SAS-050 participants: United States, Canada, Denmark, France, Germany, the Netherlands, Norway, Spain, Turkey, United Kingdom, Austria, and Sweden.
[133] Alberts and Hayes, *Campaigns of Experimentation*. 2005.
Alberts et al., *Code of Best Practice for Experimentation*. 2002.
NATO SAS-026, *Code of Best Practice for C2 Assessment*. 2002.

ments and extensions to these codes that will form the basis for future editions.

Improving the Quality and Dissemination of Data, Findings, and Instrumentation

Power to the Edge principles,[134] now embodied in DoD policy and directives,[135] include the concept and practice of moving from *smart push* to *smart pull.*[136] This shift in the approach to information dissemination is designed to promote widespread information sharing and collaboration, a necessary condition for attaining shared awareness. A cornerstone of this shift in responsibilities is the requirement for individuals and organizations, in this case, researchers, analysts, and experimenters, to post in parallel. Of course this, in and of itself, is insufficient to achieve the objective of providing users with the opportunity to shape their own information positions. This is because the information not only has to be available for users to access, but users also need to know what information is available and where and how to get it.[137] One way to accomplish this is the creation of a portal with an accompanying effort to make its existence widely known. The CCRP is in the process of creating such a portal on its Web site for C2 researchers, analysts, and experimenters. This portal will provide access and links to data, findings, and instruments of interest. However, a single portal is not in the best interests of the community, and other individuals and organizations are therefore encouraged not only to contribute to this portal, but also when appropriate to

[134] Alberts and Hayes, *Power to the Edge*. pp. 165-200.
[135] For example, the DoD Net-Centric Data Management Strategy: Metadata Registration. Memorandum by John P. Stenbit, April 3, 2003.
[136] See pages 97-99.
[137] Of course, the information is not of value unless metadata are also provided.

create similar portals on their own and provide their links to the CCRP.

Challenge of complexity

The 21st century national security environment and associated mission challenges are not just complicated, but are, in fact, complex.[138] While there is fairly widespread acknowledgement that this is the case and that we need to think differently about how we plan, organize, prepare, and carry out our missions, there is, as of yet, no clear consensus on a way ahead.

To date we have identified the need for an Information Age transformation. However, transformation has been associated with everything from modernization to disruptive concepts and capabilities. While network-centric concepts and capabilities are increasingly becoming acknowledged as important, if not critical, to transformation, we continue to make investments choices and maintain policies that undermine our ability to develop these concepts and capabilities. This is most unfortunate because network-centric concepts and capabilities, when designed and implemented according to Power to the Edge principles, are a way of coping with *complexity.*

What is complexity?[139] The word is derived from the Latin *plexus*, which means braided or entwined. *Complexus* means braided or entwined together, inseparable, or interdependent.

---

[138] A complicated system contains many components, yet displays linear, predictable behaviors. A complex system displays nonlinear unpredictable behaviors, which may in fact be unrepeatable. See: Smith, Edward A. "Effects Based Operations: The Way Ahead." Presented at the 9th ICCRTS in Copenhagen. 2004.
[139] This discussion of complexity and risk is taken from an unpublished paper: Alberts, "Complexity and Negotiation." 2005.

The implication is that a complex system cannot be deconstructed into a series of manageable or predictable pieces. The Nobel Physicist Murray Gell-Man posited this very question in a journal article[140] about a decade ago. Not surprising, he offered no simple definition for the concept but noted that "effective complexity can be high only in a region intermediate between total order and complete disorder."

Total order, a very Industrial Age concept, is linear and predictable. Changes in inputs equate to proportional changes in outputs. Outputs that correspond to the sum of two inputs are equal to the sum of the outputs arising from these individual inputs.[141] Complete disorder is randomness, and by definition impossible to predict. Complexity theory looks at the behavior of complex systems and seeks to investigate the properties and behaviors of the dynamics of nonlinear systems (whose behaviors fall between linear and random).[142]

Atkinson and Moffat[143] identify the following as the key properties of complexity, properties characteristic of "challenging" negotiations:

- nonlinear interaction,
- decentralized control,
- self-organization,
- non-equilibrium order,
- coevolution, and

---

[140] Gell-Mann, Murray. "What is Complexity?" *Complexity*. Vol 1, No 1. John Wiley and Sons. 1995.

[141] Beyerchen, Alan. "Nonlinear Science and the Unfolding of a New Intellectual Vision." *Papers in Comparative Studies*, Vol 6. Ohio State Univ. Press. 1989. p. 30.

[142] Alberts, David S. and Thomas Czerwinski. *Complexity: Global Politics, and National Security*. Washington, DC: CCRP Publication Series. 1997. Preface.

[143] Atkinson and Moffat, *The Agile Organization*. pp. 36-37.

- collectivist dynamics.

Most if not all of these are inarguably properties of 21st century military and civil-military missions. Of course, complexity would not be a problem, in and of itself, unless increasing complexity increased risk. And this is precisely the case.

Risk, from the French *risqué* and from the Italian *risco*, is the possibility of loss or injury. Thus, the level of risk is a function of the degree and nature of uncertainty and the "value" of the consequences. Even a very low probability of a catastrophic consequence is normally perceived as a high risk situation.

Complexity clearly contributes directly and indirectly to the level of risk associated with a situation. It contributes directly because it makes predictions far more uncertain and thus changes the parameters of a situation (increasing the variances of the estimated probabilities associated with specific outcomes). This makes it more difficult and risky to adopt a decision approach based on expected value and increases the uncertainty and, hence, the risk associated with any decision approach. Complexity also contributes indirectly to the level of risk associated with a situation because, in general, individuals' tolerance of uncertainty is nonlinear, with increasing uncertainty (real and/or perceived) resulting in a perception of disproportionately higher risk.

There are many reasons why the complexity of 21st century military operations increases uncertainty. Not the least of these can be found by looking at a greatly expanded and somewhat opaque effects space. For example, the direct effects that military planners traditionally consider, also known as battlefield damage assessments, are becoming less important when compared to the cascade of effects that result. Traditional measures such as loss-exchange ratios, movement of the front edge of the battle, and destruction of targets are no longer sufficient or even meaningful. The indirect effects of casualties and bomb damage are increasingly more important as they include political, social, and economic impacts that are more closely related to 21st century mission objectives.

Coping with complexity

Given that complexity increases risk in military operations, it is important that we understand what can be done to counter the adverse effects of the increased complexity that accompanies 21st century operations.

Enter **Command and Control**. The question, of course, is which **C2** Approach(es) are better suited to complex situations with their increased degree of uncertainty and increased levels of risk?

The answer is, of course, **C2** Approaches that (1) are agile and (2) take full advantage of all of the available information and assets. The second of these enables us to reduce uncertainties when and where we can, while the first allows us to deal with the residual uncertainty.

There is a growing body of evidence that supports the hypothesis that network-centric **C2** Approaches that embody Power to the Edge principles both can get the most from the information and assets that are available and are more agile than traditional **C2** Approaches. Given that we have good reason to believe that new, network-centric, Power to the Edge **C2**

Approaches can counter the adverse impacts of increased complexity, exploring these approaches should be a high priority.

FINAL THOUGHTS

Command and Control has the reputation of being an arcane subject. While this has always been unfortunate, given the central role that **C2** needs to play in our transformation to effects-based Network Centric Operations, a failure to create increased awareness about why new **C2** Approaches are needed as well as the nature of these new approaches will seriously hamper progress. Understanding **C2** is not about how it has been accomplished in the past. Rather, it is about what functions of command and what functions of control need to be accomplished. It is about potentially useful **Command and Control** Approaches and the value propositions that trace improvements in **C2** to measures of effectiveness. It is also about creating more awareness of the growing importance of agility as a primary measure of effectiveness and the link between agile **C2** and agile organizations and operations. Finally, interest in and concern about **Command and Control** should no longer be left to the "**C2** specialist." It is far too important. **Command and Control** is everyone's concern.

ABOUT THE AUTHORS

DR. DAVID S. ALBERTS

Dr. Alberts is currently the Director of Research for the Office of the Assistant Secretary of Defense (Networks and Information Integration). One of his principal responsibilities is DoD's Command and Control Research Program, a program whose mission is to develop a better understanding of the implications of the Information Age for national security and Command and Control. Prior to this he was the Director, Advanced Concepts, Technologies, and Information Strategies (ACTIS) and Deputy Director of the Institute for National Strategic Studies at the National Defense University. Dr. Alberts was also responsible for managing the Center for Advanced Concepts and Technology (ACT) and the School of Information Warfare and Strategy (SIWS).

Dr. Alberts is credited with significant contributions to our understanding of the Information Age and its implications for national security, military Command and Control, and organization. These include the tenets of Network Centric Warfare, the coevolution of mission capability packages, new approaches to Command and Control, evolutionary acquisition of Command and Control systems, the nature and conduct of Information Warfare, and most recently the "Power to the Edge" principles that currently guide efforts to provide the enabling "infostructure" of DoD transformation, and efforts to better understand Edge organizations and the nature of Command and Control in a networked environment. These works have developed a following that extends well beyond the DoD and the Defense Industry.

His contributions to the conceptual foundations of Defense transformation are complemented by more than 25 years of pragmatic experience developing and introducing leading edge technology into private and public sector organizations. This extensive applied experience is augmented by a distinguished academic career in Computer Science and Operations Research and by Government service in senior policy and management positions.

Dr. Richard E. Hayes

As President and founder of Evidence Based Research, Inc., Dr. Hayes specializes in multi-disciplinary analyses of Command and Control, intelligence, and national security issues; the identification of opportunities to improve support to decisionmakers in the defense and intelligence communities; the design and development of systems to provide that support; and the criticism, test, and evaluation of systems and procedures that provide such support. His areas of expertise include crisis management; political-military issues; research methods; experimental design; simulation and modeling; test and evaluation; military command, control, communication, and intelligence (C3I or NII); and decision-aiding systems. Since coming to Washington in 1974, Dr. Hayes has established himself as a leader in bringing the systematic use of evidence and the knowledge base of the social sciences into play in support of decisionmakers in the national security community, domestic agencies, and major corporations. He has initiated several programs of research and lines of business that achieved national attention and many others that directly influenced policy development in client organizations.

Dr. Hayes has co-authored several other CCRP titles, including: *Campaigns of Experimentation* (2005), *Power to the Edge* (2003), *The Code of Best Practice for Experimentation* (2002), *Understanding Information Age Warfare* (2001), and *Command Arrangements for Peace Operations* (1995).

BIBLIOGRAPHY

"NATO-Russia Glossary of Contemporary Political and Military Terms." May 2002. http://www.nato.int/docu/glossary/eng/15-main.pdf (April 2005)

"NCO Conceptual Framework Version 2.0." Prepared by Evidence Based Research, Inc. Vienna, VA. 2004.

"SCUDHunt and Shared Situation Awareness." Experiment conducted by ThoughtLink and the Center for Naval Analyses for the Defense Advanced Research Projects Agency (DARPA). http://www.thoughtlink.com/wae.htm (Oct 2005).

Ackoff, Russel L. "From Data to Wisdom." *Journal of Applies Systems Analysis*. Vol 16. 1989.

Ackoff, Russel L. and Maurice W. Sasieni. *Fundamentals of Operations Research*. New York, London, Sidney: John Wiley & Sons, Inc. 1968.

Albert, Reka and Albert-Laszlo Barabasi. "Statistical Mechanics of Complex Networks." *Reviews of Modern Physics*. Vol 74, No 1. 2002. www.cs.unibo.it/~babaoglu/courses/cas/tutorials/rmp.pdf (March 2006)

Alberts, David S. "Complexity and Negotiation." Unpublished work. 2005.

Alberts, David S. "Mission Capability Packages." National Defence University: Strategic Forum. Institute for National Strategic Studies. Number 14. 1995.

Alberts, David S. and Richard E. Hayes. *Campaigns of Experimentation.* Washington, DC: CCRP Publication Series. 2002.

Alberts, David S. and Richard E. Hayes. *Command Arrangements for Peace Operations.* Washington, DC: CCRP Publication Series. 1995. http://www.dodccrp.org/publications/pdf/Alberts_Arrangements.pdf (March 2006)

Alberts, David S. and Richard E. Hayes. *Power to the Edge: Command and Control in the Information Age.* Washington, DC: CCRP Publication Series. 2003. http://www.dodccrp.org/publications/pdf/Alberts_Power.pdf (March 2006)

Alberts, David S. and Thomas Czerwinski. *Complexity: Global Politics, and National Security.* Washington, DC: CCRP Publication Series. 1997. http://www.dodccrp.org/publications/pdf/Alberts_Complexity_Global.pdf (March 2006)

Alberts, David S. *Information Age Transformation: Getting to a 21st Century Military.* Washington, DC: CCRP Publication Series. 2002. http://www.dodccrp.org/publications/pdf/Alberts_IAT.pdf (March 2006)

Alberts, David S. *The Unintended Consequences of the Information Age Technologies.* National Defence University: Institute for National Strategic Studies. 1996. http://www.dodccrp.org/publications/pdf/Alberts_Unintended.pdf (March 2006)

Alberts, David S., John J. Garstka, and Frederick P. Stein. *Network Centric Warfare: Developing and Leveraging Information Superiority.* Washington, DC: CCRP Publication Series. 1999. http://www.dodccrp.org/publications/pdf/Alberts_UIAW.pdf (March 2006)

Alberts, David S., John J. Garstka, Richard E. Hayes, and David T. Signori. *Understanding Information Age Warfare.* Washington, DC: CCRP Publication Series. 2001.

http://www.dodccrp.org/publications/pdf/Alberts_UIAW.pdf (March 2006)

Alberts, David S., Richard E. Hayes, Dennis K. Leedom, John E. Kirzl, and Daniel T. Maxwell. *Code of Best Practice for Experimentation.* Washington, DC: CCRP Publication Series. 2002. http://www.dodccrp.org/publications/pdf/Alberts_Experimentation.pdf (March 2006)

Atkinson, Simon Reay and James Moffat. *The Agile Organization: From Information Networks to Complex Effects and Agility.* Washington, DC: CCRP Publications Series. 2005. http://www.dodccrp.org/publications/pdf/Atkinson_Agile.pdf (March 2006)

Barabasi, Albert-Laszlo. *Linked: How Everything Is Connected to Everything Else and What It Means.* New York, NY: Penguin Group. 2003.

Beyerchen, Alan. "Nonlinear Science and the Unfolding of a New Intellectual Vision." Papers in Comparative Studies, Vol 6. Ohio State University Press. 1989.

Booz Allen Hamilton. "Network Centric Operations Case Study: Naval Special Warfare Group One (NSWG-1)." 2004. (Unpublished at this time).

Brown, Rupert. *Group Processes: Dynamics Within and Between Groups.* Malden, MA: Blackwell Publishing. 2000.

Buchanan, Mark. *Nexus: Small Worlds and the Groundbreaking Science of Networks.* New York, NY: W.W. Norton & Company, Inc. 2002.

Chase, W.G. and H.A. Simon. "Perception in Chess." *Cognitive Psychology.* No 4. 1973.

Comfort, Louise K. "Risk, Security, and Disaster Management." *Annual Review of Political Science.* Vol 8. 2005.

Cray, Ed. *General of the Army, George C. Marshall: Soldier and Statesman.* New York, NY: Cooper Square Press. 2000.

Defense Technical Information Center. DoD Dictionary of Military and Associated Terms. Joint Publication 1-02. http://www.dtic.mil/doctrine/jel/doddict/data/c/01105.html (April 2005)

Degenne, Alain and Michael Forsé. *Introducing Social Networks.* Thousand Oaks, CA: Sage Publications, Ltd. 1999.

Department of Defense, Chairman of the Joint Chiefs of Staff. *Joint Vision 2010.* Washington, DC: GPO. 1996.

Department of Defense, Chairman of the Joint Chiefs of Staff. *Joint Vision 2020.* Washington, DC: GPO. 2000.

Department of Defense, CIO. Information Management Directorate. "Communities of Interest in Net-Centric DoD, Version 1.0." May 19, 2004.

Department of Defense. "Quarterly Defense Review Report." September 30, 2001

DoD Net-Centric Data Management Strategy: Metadata Registration. Memorandum by John P. Stenbit, April 3, 2003. http://www.defenselink.mil/nii/doc/DMmemo20030403.pdf (Oct 2005)

Druzhinin, V. V. and D. S. Kontorov. *Decision Making and Automation: Concept, Algorithm, Decision (A Soviet View).* Moscow, Russia: CCCP Military Publishing. 1972.

Filiberti, Edward J. *How the Army Runs: A Senior Leader Reference Handbook, 2003-2004.* Carlisle, PA; U. S. Army War College Department of Command, Leadership and Management, 2003. http://www.carlisle.army.mil/usawc/dclm/htar.htm (April 2005)

Fisher, B. Aubrey. *Small Group Decision Making: Communication and the Group Process.* New York, NY: McGraw-Hill Book Company. 1980.

Galdorisi, George, Jeff Grossman, Mike Reilly, Jeff Clarkson, and Chris Priebe. "Composeable FORCEnet Command and Control: The Key to Energizing the Global Information Grid to Enable Superior Decision Making." Presented at the 2004 Command and Control

Research Technology Symposium. San Diego, CA. June 15-17, 2004.
http://www.dodccrp.org/events/2004/CCRTS_San_Diego/ CD/track08.htm
(March 2005)

Gell-Mann, Murray. "What is Complexity?" *Complexity.* Vol 1, No 1. John Wiley and Sons. 1995.

Gladwell, Malcolm. *The Tipping Point: How Little Things Can Make a Big Difference.* United States: Little, Brown and Company. 2000.

Handley, Holly A.H. "Adaptive Architecture for Command and Control and the Model Driven Experimentation Paradigm."
http://viking.gmu.edu/a2c2/a2c2.htm
(April 2005)

Hayes, Richard E. "Systematic Assessment of C2 Effectiveness and Its Determinants." Presented at the 1994 Symposium on Command and Control Research and Decision Aids. Monterey, CA. June 21-23, 1994.

Hayes, Richard E., Conrad Strack, and Daniel Bucioni. "Headquarters Effectiveness Program Summary." McLean, VA: Defense Systems, Inc. 1983.

Hayes, Richard E., Mark Hainline, Conrad Strack, and Daniel Bucioni. "Theater Headquarters Effectiveness: Its Measurement and Relationship to Size Structure, Functions, and Linkage." McLean, VA: Defense Systems, Inc. 1983.

Honegger, Barbara. "NPS IT Fly-in Team Reconnects Tsunami Survivors to the World." Naval Postgraduate School. Feb 9, 2005.
http://www.nps.edu/PAO/ArchiveDetail.aspx?id=1091&role=pao&area=news
(April 2005)

Janis, Irving. *Groupthink: Psychological Studies of Policy Decisions and Fiascoes.* Boston, MA: Houghton Mifflin. 1982.

Johnson, Steven. *Emergence: The Connected Lives of Ants, Brains, Cities, and Software.* New York, NY: Touchstone. 2001.

Kahneman, Daniel and Amos Tversky. "Prospect Theory: An Analysis of Decision under Risk." *Econometrica*. Vol 47, Iss 2. March 1979.

Kelly, Terrence K. "Transformation and Homeland Security: Dual Challenges for the U.S. Army." *Parameters*. Carlisle, PA: U.S. Army War College. Summer 2003.

Khalilzad, Zalmay M. and David A. Ochmanek. *Strategy and Defense Planning for the 21st Century*. Santa Monica, CA: Rand. 1997.

Kuhn, Thomas. *The Structure of Scientific Revolutions*. Chicago, IL: University of Chicago Press. 1996.

Merriam-Webster Online Thesaurus. http://www.merriam-webster.com/cgi-bin/mwthesadu (April 03, 2005)

Milgram, Stanley. "The small world problem." *Psychology Today*. Vol 2. 1967.

NATO. *Code of Best Practices for C2 Assessment*. Washington, DC: CCRP Publication Series. 2002. http://www.dodccrp.org/publications/pdf/NATO_COBP.pdf (March 2006)

NATO RTO. "System, Analysis, and Studies (SAS)." http://www.rta.nato.int/Main.asp?topic=sas.htm (April 2005)

NATO SAS-050. "SAS-050 Conceptual Model Version 1.0." http://www.dodccrp.org/SAS/SAS-050%20Final%20Report.pdf (March 2006)

Network Centric Warfare Department of Defense Report to Congress. Washington DC. July 2001. http://www.dodccrp.org/research/ncw/ncw_report/flash.htm (March 2006)

Noble, David F. "Understanding and Applying the Cognitive Foundations of Effective Teamwork." Prepared for the Office of Naval Research by Evidence Based Research, Inc. April 23, 2004.

Office of Force Transformation. "NCO Conceptual Framework, Version 1." Prepared by Evidence Based Research, Inc. Vienna, VA. 2003. http://www.oft.osd.mil/library/library_files/document_353_NCO%20CF%20Version%201.0%20(FINAL).d oc (January 2006)

Office of Force Transformation. "NCO Conceptual Framework, Version 2." Prepared by Evidence Based Research, Inc. Vienna, VA. 2004.

Owen, Don and Eugene K. Hopkins. "Southeast Asia Tsunami Research Project." Report commissioned by the CCRP and prepared by Evidence Based Research, Inc. 2005.

Pigeau, Ross and Carol McCann. "Re-conceptualizing Command and Control." *Canadian Military Journal*. Vol 3, No. 1. Spring 2002.

Prins, Gwyn. *The Heart of War: On Power, Conflict, and Obligation in the 21st Century*. London, UK: Routledge. 2002.

Saariluoma, P. "Chess Players Recall of Auditory Presented Chess Positions." *European Journal of Cognitive Psychology*. No 1. 1989.

Scott, John. *Social Network Analysis: A Handbook*. London, UK: Sage Publications, Ltd. 2000.

Smith, Edward A. "Effects Based Operations: The Way Ahead." Presented at the 9th ICCRTS in Copenhagen. 2004. http://www.au.af.mil/au/awc/awcgate/ccrp/ ebo_way_ahead.pdf (March 2006)

Smith, Edward A. *Effects Based Operations: Applying Network Centric Warfare in Peace, Crisis, and War*. Washington, DC: CCRP Publication Series. 2002. http://www.dodccrp.org/publications/pdf/Smith_EBO.PDF (March 2006)

Tsouras, Peter G. *The Greenhill Dictionary of Military Quotations*. London, UK: Greenhill Books. 2004.

Wasserman, Stanley and Katherine Faust. *Social Network Analysis: Methods and Application*. Cambridge, UK: Cambridge University Press. 1999.

Watts, Duncan J. *Small Worlds: The Dynamics of Networks between Order and Randomness*. Princeton, NJ: W.W. Princeton University Press. 1999.

Weick, K.E. & K.M. Sutcliffe. *Managing the Unexpected: Assuring High Performance in an Age of Complexity.* San Francisco, CA: Jossey-Bass. 2001.

Zeleny, M. "Management Support Systems: Towards Integrated Knowledge Management." *Human Systems Management*. Vol 7. 1987.

Index

L

M

N

O

P

Q

R

S

Catalog of CCRP Publications

(* denotes a title no longer available in print)

Coalition Command and Control*
(Maurer, 1994)

Peace operations differ in significant ways from traditional combat missions. As a result of these unique characteristics, command arrangements become far more complex. The stress on command and control arrangements and systems is further exacerbated by the mission's increased political sensitivity.

The Mesh and the Net
(Libicki, 1994)

Considers the continuous revolution in information technology as it can be applied to warfare in terms of capturing more information (mesh) and how people and their machines can be connected (net).

Command Arrangements for Peace Operations
(Alberts & Hayes, 1995)

By almost any measure, the U.S. experience shows that traditional C2 concepts, approaches, and doctrine are not particularly well suited for peace operations. This book (1) explores the reasons for this, (2) examines alternative command arrangement approaches, and (3) describes the attributes of effective command arrangements.

Standards: The Rough Road to the Common Byte
(Libicki, 1995)

The inability of computers to "talk" to one another is a major problem, especially for today's high technology military forces. This study by the Center for Advanced Command Concepts and Technology looks at the growing but confusing body of information technology standards. Among other problems, it discovers a persistent divergence between the perspectives of the commercial user and those of the government.

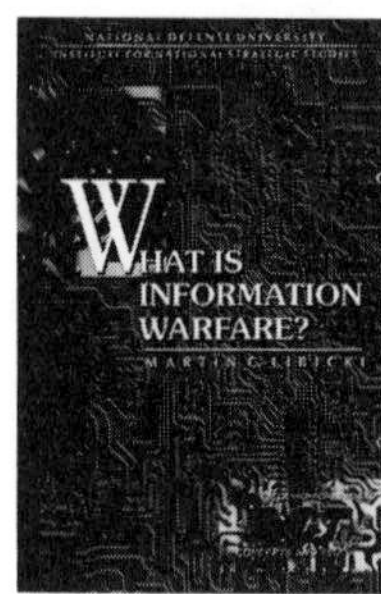

What Is Information Warfare?*
(Libicki, 1995)

Is Information Warfare a nascent, perhaps embryonic art, or simply the newest version of a time-honored feature of warfare? Is it a new form of conflict that owes its existence to the burgeoning global information infrastructure, or an old one whose origin lies in the wetware of the human brain but has been given new life by the Information Age? Is it a unified field or opportunistic assemblage?

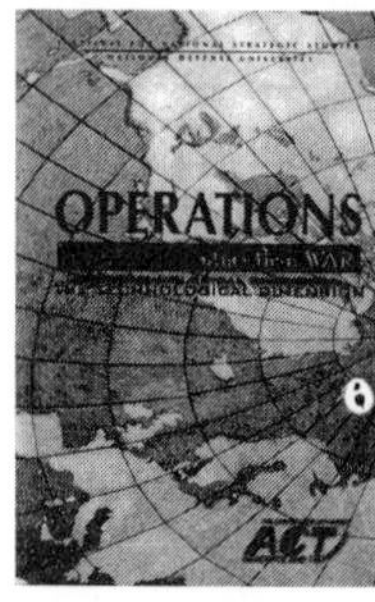

Operations Other Than War*
(Alberts & Hayes, 1995)

This report documents the fourth in a series of workshops and roundtables organized by the INSS Center for Advanced Concepts and Technology (ACT). The workshop sought insights into the process of determining what technologies are required for OOTW. The group also examined the complexities of introducing relevant technologies and discussed general and specific OOTW technologies and devices.

Dominant Battlespace Knowledge*
(Johnson & Libicki, 1996)

The papers collected here address the most critical aspects of that problem—to wit: If the United States develops the means to acquire dominant battlespace knowledge, how might that affect the way it goes to war, the circumstances under which force can and will be used, the purposes for its employment, and the resulting alterations of the global geomilitary environment?

Interagency and Political-Military Dimensions of Peace Operations: Haiti - A Case Study
(Hayes & Wheatley, 1996)

This report documents the fifth in a series of workshops and roundtables organized by the INSS Center for Advanced Concepts and Technology (ACT). Widely regarded as an operation that "went right," Haiti offered an opportunity to explore interagency relations in an operation close to home that had high visibility and a greater degree of interagency civilian-military coordination and planning than the other operations examined to date.

The Unintended Consequences of the Information Age*
(Alberts, 1996)

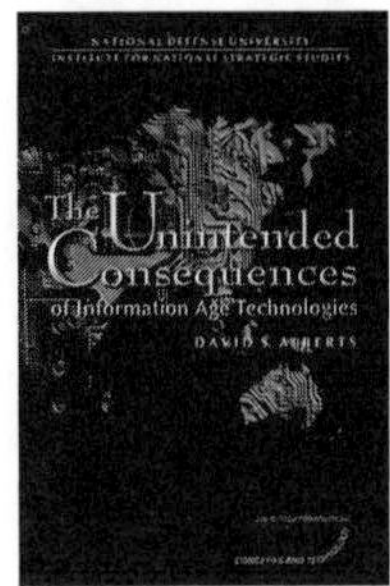

The purpose of this analysis is to identify a strategy for introducing and using Information Age technologies that accomplishes two things: first, the identification and avoidance of adverse unintended consequences associated with the introduction and utilization of information technologies; and second, the ability to recognize and capitalize on unexpected opportunities.

Joint Training for Information Managers*
(Maxwell, 1996)

This book proposes new ideas about joint training for information managers over Command, Control, Communications, Computers, and Intelligence (C4I) tactical and strategic levels. It suggests a substantially new way to approach the training of future communicators, grounding its argument in the realities of the fast-moving C4I technology.

Defensive Information Warfare*
(Alberts, 1996)

This overview of defensive information warfare is the result of an effort, undertaken at the request of the Deputy Secretary of Defense, to provide background material to participants in a series of interagency meetings to explore the nature of the problem and to identify areas of potential collaboration.

Command, Control, and the Common Defense
(Allard, 1996)

The author provides an unparalleled basis for assessing where we are and were we must go if we are to solve the joint and combined command and control challenges facing the U.S. military as it transitions into the 21st century.

Shock & Awe: Achieving Rapid Dominance*
(Ullman & Wade, 1996)

The purpose of this book is to explore alternative concepts for structuring mission capability packages around which future U. S. military forces might be configured.

Information Age Anthology: Volume I*
(Alberts & Papp, 1997)

In this first volume, we will examine some of the broader issues of the Information Age: what the Information Age is; how it affects commerce, business, and service; what it means for the government and the military; and how it affects international actors and the international system.

Complexity, Global Politics, and National Security*
(Alberts & Czerwinski, 1997)

The charge given by the President of the National Defense University and RAND leadership was threefold: (1) push the envelope; (2) emphasize the policy and strategic dimensions of national defense with the implications for complexity theory; and (3) get the best talent available in academe.

Target Bosnia: Integrating Information Activities in Peace Operations*
(Siegel, 1998)

This book examines the place of PI and PSYOP in peace operations through the prism of NATO operations in Bosnia-Herzegovina.

Coping with the Bounds
(Czerwinski, 1998)

The theme of this work is that conventional, or linear, analysis alone is not sufficient to cope with today's and tomorrow's problems, just as it was not capable of solving yesterday's. Its aim is to convince us to augment our efforts with nonlinear insights, and its hope is to provide a basic understanding of what that involves.

Information Warfare and International Law*
(Greenberg, Goodman, & Soo Hoo, 1998)

The authors, members of the Project on Information Technology and International Security at Stanford University's Center for International Security and Arms Control, have surfaced and explored some profound issues that will shape the legal context within which information warfare may be waged and national information power exerted in the coming years.

Lessons From Bosnia: The IFOR Experience*
(Wentz, 1998)

This book tells the story of the challenges faced and innovative actions taken by NATO and U.S. personnel to ensure that IFOR and Operation Joint Endeavor were military successes. A coherent C4ISR lessons learned story has been pieced together from firsthand experiences, interviews of key personnel, focused research, and analysis of lessons learned reports provided to the National Defense University team.

Doing Windows: Non-Traditional Military Responses to Complex Emergencies
(Hayes & Sands, 1999)

This book provides the final results of a project sponsored by the Joint Warfare Analysis Center. Our primary objective in this project was to examine how military operations can support the long-term objective of achieving civil stability and durable peace in states embroiled in complex emergencies.

Network Centric Warfare
(Alberts, Garstka, & Stein, 1999)

It is hoped that this book will contribute to the preparations for NCW in two ways. First, by articulating the nature of the characteristics of Network Centric Warfare. Second, by suggesting a process for developing mission capability packages designed to transform NCW concepts into operational capabilities.

Behind the Wizard's Curtain
(Krygiel, 1999)

There is still much to do and more to learn and understand about developing and fielding an effective and durable infostructure as a foundation for the 21st century. Without successfully fielding systems of systems, we will not be able to implement emerging concepts in adaptive and agile command and control, nor will we reap the potential benefits of Network Centric Warfare.

Confrontation Analysis: How to Win Operations Other Than War
(Howard, 1999)

A peace operations campaign (or operation other than war) should be seen as a linked sequence of confrontations, in contrast to a traditional, warfighting campaign, which is a linked sequence of battles. The objective in each confrontation is to bring about certain "compliant" behavior on the part of other parties, until in the end the campaign objective is reached. This is a state of sufficient compliance to enable the military to leave the theater.

Information Campaigns for Peace Operations
(Avruch, Narel, & Siegel, 2000)

In its broadest sense, this report asks whether the notion of struggles for control over information identifiable in situations of conflict also has relevance for situations of third-party conflict management—for peace operations.

Information Age Anthology: Volume II*
(Alberts & Papp, 2000)

Is the Information Age bringing with it new challenges and threats, and if so, what are they? What sorts of dangers will these challenges and threats present? From where will they (and do they) come? Is information warfare a reality? This publication, Volume II of the Information Age Anthology, explores these questions and provides preliminary answers to some of them.

Information Age Anthology: Volume III*
(Alberts & Papp, 2001)

In what ways will wars and the military that fight them be different in the Information Age than in earlier ages? What will this mean for the U.S. military? In this third volume of the Information Age Anthology, we turn finally to the task of exploring answers to these simply stated, but vexing questions that provided the impetus for the first two volumes of the Information Age Anthology.

Understanding Information Age Warfare
(Alberts, Garstka, Hayes, & Signori, 2001)

This book presents an alternative to the deterministic and linear strategies of the planning modernization that are now an artifact of the Industrial Age. The approach being advocated here begins with the premise that adaptation to the Information Age centers around the ability of an organization or an individual to utilize information.

Information Age Transformation
(Alberts, 2002)

This book is the first in a new series of CCRP books that will focus on the Information Age transformation of the Department of Defense. Accordingly, it deals with the issues associated with a very large governmental institution, a set of formidable impediments, both internal and external, and the nature of the changes being brought about by Information Age concepts and technologies.

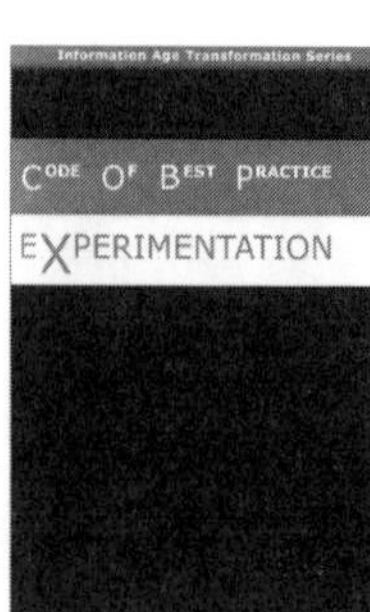

Code of Best Practice for Experimentation
(CCRP, 2002)

Experimentation is the lynch pin in the DoD's strategy for transformation. Without a properly focused, well-balanced, rigorously designed, and expertly conducted program of experimentation, the DoD will not be able to take full advantage of the opportunities that Information Age concepts and technologies offer.

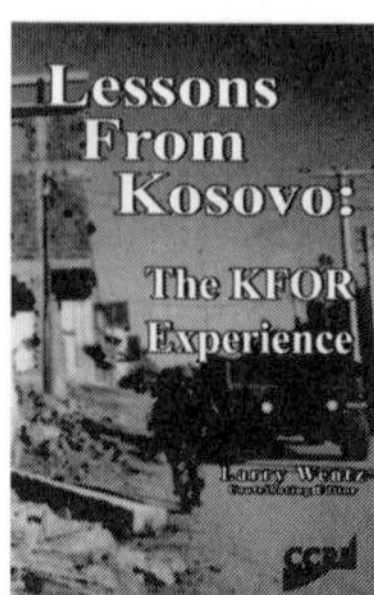

Lessons From Kosovo: The KFOR Experience
(Wentz, 2002)

Kosovo offered another unique opportunity for CCRP to conduct additional coalition C4ISR-focused research in the areas of coalition command and control, civil-military cooperation, information assurance, C4ISR interoperability, and information operations.